## Understanding Natural Selection

Natural selection, as introduced by Charles Darwin in the *Origin of Species* (1859), has always been a topic of great conceptual and empirical interest. This book puts Darwin's theory of evolution in historical context showing that, in important respects, his central mechanism of natural selection gives the clue to understanding the nature of organisms. Natural selection has important implications, not just for the understanding of life's history – single-celled organisms to humans – but also for our understanding of contemporary social norms, as well as the nature of religious belief. The book is written in clear, non-technical language, appealing not just to philosophers, historians, and biologists, but also to general readers who find thinking about important issues both challenging and exciting.

Michael Ruse is Professor Emeritus at the University of Guelph, Ontario. He is a Guggenheim Fellow, a Gifford Lecturer, a Fellow of the Royal Society of Canada, and the recipient of four honorary degrees. He is the author/editor of over sixty books, including *The Cambridge Encylopedia of Darwin and Evolutionary Thought* (Cambridge University Press, 2013); *The Gaia Hypothesis: Life on a Pagan Planet* (University of Chicago Press, 2013); *Darwinism as Religion: What Literature Tells Us about Evolution* (Oxford University Press, 2016), and *The Cambridge History of Atheism* (Cambridge University Press, 2021).

T0054594

The **Understanding Life** series is for anyone wanting an engaging and concise way into a key biological topic. Offering a multidisciplinary perspective, these accessible guides address common misconceptions and misunderstandings in a thoughtful way to help stimulate debate and encourage a more in-depth understanding. Written by leading thinkers in each field, these books are for anyone wanting an expert overview that will enable clearer thinking on each topic.

Series Editor: Kostas Kampourakis http://kampourakis.com

**Published titles:**

| | | |
|---|---|---|
| *Understanding Evolution* | Kostas Kampourakis | 9781108746083 |
| *Understanding Coronavirus* | Raul Rabadan | 9781108826716 |
| *Understanding Development* | Alessandro Minelli | 9781108799232 |
| *Understanding Evo-Devo* | Wallace Arthur | 9781108819466 |
| *Understanding Genes* | Kostas Kampourakis | 9781108812825 |
| *Understanding DNA Ancestry* | Sheldon Krimsky | 9781108816038 |
| *Understanding Intelligence* | Ken Richardson | 9781108940368 |
| *Understanding Metaphors in the Life Sciences* | Andrew S. Reynolds | 9781108940498 |
| *Understanding Cancer* | Robin Hesketh | 9781009005999 |
| *Understanding How Science Explains the World* | Kevin McCain | 9781108995504 |
| *Understanding Race* | Rob DeSalle and Ian Tattersall | 9781009055581 |
| *Understanding Human Evolution* | Ian Tattersall | 9781009101998 |
| *Understanding Human Metabolism* | Keith N. Frayn | 9781009108522 |
| *Understanding Fertility* | Gab Kovacs | 9781009054164 |
| *Understanding Forensic DNA* | Suzanne Bell and John M. Butler | 9781009044011 |
| *Understanding Natural Selection* | Michael Ruse | 9781009088329 |

**Forthcoming:**

| | | |
|---|---|---|
| *Understanding Life in the Universe* | Wallace Arthur | 9781009207324 |
| *Understanding Species* | John S. Wilkins | 9781108987196 |
| *Understanding Creationism* | Glenn Branch | 9781108927505 |
| *Understanding the Nature–Nurture Debate* | Eric Turkheimer | 9781108958165 |

# Understanding Natural Selection

MICHAEL RUSE
University of Guelph

CAMBRIDGE
UNIVERSITY PRESS

## CAMBRIDGE
### UNIVERSITY PRESS

University Printing House, Cambridge CB2 8BS, United Kingdom

One Liberty Plaza, 20th Floor, New York, NY 10006, USA

477 Williamstown Road, Port Melbourne, VIC 3207, Australia

314–321, 3rd Floor, Plot 3, Splendor Forum, Jasola District Centre,
New Delhi – 110025, India

103 Penang Road, #05–06/07, Visioncrest Commercial, Singapore 238467

Cambridge University Press is part of the University of Cambridge.

It furthers the University's mission by disseminating knowledge in the pursuit of
education, learning, and research at the highest international levels of excellence.

www.cambridge.org
Information on this title: www.cambridge.org/9781316514788
DOI: 10.1017/9781009090865

First published 2023

Printed in the United Kingdom by TJ Books Limited, Padstow Cornwall

*A catalogue record for this publication is available from the British Library.*

ISBN 978-1-316-51478-8 Hardback
ISBN 978-1-009-08832-9 Paperback

"In this brief book written for the general reader, Michael Ruse skillfully weaves together the history and philosophy of science to explore natural selection, the concept at the heart of Darwin's celebrated theory of evolution. The writing is brisk, engaging, thoughtful and at times fun, typical of the kind of work we have come to expect from someone who has a devoted a lifetime of study to understanding Darwin and his theory."

> Vassiliki Betty Smocovitis, Professor of the History of
> Science, University of Florida, USA

"Natural selection is one of the most important and contested ideas in modern science, helping us understand much of the functional design and order we observe in living nature. In his inimitable way, Michael Ruse gives the definitive account of natural selection, from its Darwinian origins and meta-phorical foundation to the many historical, philosophical and scientific controversies that have swirled about it in the last century and a half. If you want to understand natural selection, you can do no better than a careful reading of this compact, highly informative and lively book. It is truly a tour de force."

> Richard A. Richards, Professor and Chair, Department of
> Philosophy, University of Alabama, USA

For Kenneth and Christa Dorter

# Contents

# Foreword

If you are interested in this book, you have probably heard about natural selection and Darwin before. The process is often represented as a very simple one: some individuals make it, others do not; some survive and reproduce, whereas others die before leaving any offspring. Therefore, only some will pass on their DNA and their biological features to the next generation. If this is done for a long time, the population may evolve to a different one; to take a widely known example, if giraffes with longer necks have an advantage in browsing leaves from trees compared to those with shorter necks, and therefore are more successful in surviving and reproducing, then the giraffe population may evolve to one with a higher average neck length. The long-necked giraffes will thus be naturally selected. If it is that simple, then why a whole book on natural selection? As Michael Ruse shows in this wonderful tour-de-force, nothing has never been simple regarding natural selection. Starting from the writings of Charles Darwin, Ruse takes us on a magnificent journey through history, science, literature, politics, and more during the last 200 years or so. Along the way, he shows that the importance of natural selection has repeatedly been questioned and has been misunderstood countless times. Through his unparalleled knowledge of history and philosophy, Ruse makes the case for the importance of natural selection in evolutionary theory. Understanding natural selection is essential for understanding evolutionary theory, as well as a critical step towards understanding how we, humans, are a contingent outcome of evolution, but also one of its triumphs.

**Kostas Kampourakis, Series Editor**

# Preface

This is a book on the concept of natural selection, the chief cause of evolutionary change, first introduced and explained in detail by Charles Darwin, in 1859, in his *Origin of Species*. It is, if anything, a cause even more cherished and highlighted by professional evolutionists today. I have just retired after fifty-five years of teaching philosophy, with much of my effort devoted to issues arising from the biological sciences, especially the work of Charles Darwin in both a historical and a philosophical context. Along the way, I have written or edited many books, also often directed to Darwinian themes. Truly, my greatest joy is when I can bring together my role as an educator and as a scholar. To be included in this series on biology, edited by Kostas Kampourakis, makes me feel incredibly pleased and tremendously honored. The intent of the series is to produce books uncompromising in scholarship, and yet, with no condescension, readable and comprehensible by all who want to learn about the topics under discussion. Looking at earlier contributions to the series, I know that I have high standards to emulate.

Apart from my deepest thanks to my editor, above all I want to thank my students, recent and long in the past, graduate and undergraduate, young and old, male and female, clever and those falling into the category that, euphemistically, educators refer to as "late developers." I cannot say it was always fun – anyone who says that a logic class is a bundle of laughs is a liar – but it was always worthwhile, and we on both sides of the classroom realized that. Closer to home, I cannot thank my wife Lizzie enough for her love and support. And no acknowledgments would be complete without mention of my cairn terriers,

Scruffy McGruff and Duncan Donut. They are ever ready to make me down tools and go for a walk. They show that humans are but a very small part of the story.

My dedication is to two dear friends of very long standing. Thank you, natural selection, for producing people such as these.

# 1 The Origin of Species

On November 24, 1859, the English naturalist Charles Robert Darwin published *On the Origin of Species by Means of Natural Selection, or the Preservation of Favoured Races in the Struggle for Life*. In that book, he argued that all organisms, living and dead, were produced by a long, slow, natural process, from a very few original organisms. He called the process "natural selection," later giving it the alternative name of "the survival of the fittest." This first chapter is devoted to presenting (without critical comment) the argument of the *Origin*, very much with an eye to the place and role of natural selection. As a preliminary, it should be noted that the *Origin*, for all it is one of the landmark works in the history of science, was written in a remarkably "user-friendly" manner. It is not technical, the arguments are straightforward, the illustrative examples are relevant and easy to grasp, the mathematics is at a minimum, meaning non-existent. Do not be deceived. The *Origin* is also a very carefully structured piece of work. Darwin knew exactly what he was doing when he set pen to paper. It is easiest to regard the *Origin* as being (as Julius Caesar told us), like Gaul, divided into three parts: artificial selection, natural selection, consequences. We shall be guided by this division. Although Darwin was to describe the *Origin* as "one long argument," later we shall see reason to think that Darwin himself would have been comfortable with our triune approach (Fig. 1.1).

A linguistic point as we plunge in. The term "evolution" had traditionally referred to the change of organisms as they individually develop: that is, "ontogeny." It was only in the 1850s that the term came commonly to refer to change of organisms through time: "phylogeny". Darwin had no strong

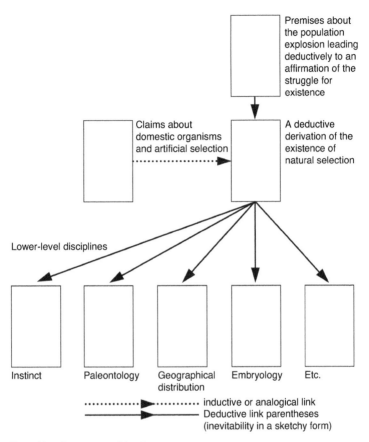

**Figure 1.1**   The structure of the *Origin*.

objection to the term, and indeed the last word of the *Origin* is "evolved." In later works he used it constantly. Usually in the *Origin*, though, he used terms like "descent with modification." Somewhat anachronistically, since "evolution" is the term we use today, this is the term I shall use in this book.

## Artificial Selection

The Industrial Revolution occurred in Britain in the second half of the eighteenth century and the first half of the nineteenth. Machines started to take over what previously had been done by hand, and this meant that people left the land and moved to cities, where factories housed the power-driven weaving looms producing cloth, where the furnaces for making metal goods blazed away, and where cheap mass-produced pottery could be fashioned from incoming loads of clay and other raw materials. But, to have an industrial revolution, you must at the same time have an agricultural revolution – more food for those in the factories, produced by fewer workers on the farms. It was realized that a key to such a revolution was selective breeding – fatter pigs, beefier cattle, shaggier sheep (Fig. 1.2 shows a pig bred for the market). Darwin came from Shropshire, in the heart of agricultural England, and he knew from the first that the key to success was picking the better specimens and breeding from them. It was almost preordained, therefore, that when his thoughts turned to evolution – organisms coming naturalistically from earlier, simpler forms – he was led first to human-driven organic change and its reasons. No surprise therefore that Chapter One of the *Origin* was devoted to selective breeding, both in the farmyard and by the fancier – pigeons, songbirds, fighting dogs.

**Figure 1.2**   A fat pig bred for bacon.

Charles Darwin was ever a Lamarckian, meaning that (after the mechanism supposed by the French evolutionist Jean Baptiste de Lamarck) he accepted that change could come directly from forces on the individual organism; notably, he believed in the inheritance of acquired characteristics, where something like the blacksmith's strong arms, the result of a lifetime at the forge, could be passed on directly to offspring. Lamarckism notwithstanding, it was selection that was the key force of change. Darwin focused in on pigeons, long a favorite of breeders, which came in many different varieties, some enough (by their looks) to be considered separate species. The key phrase is "by their looks." Darwin thought that they came from the same source and were not indeed different species because they could still interbreed. What, then, had wrought the differences? "The key is man's power of accumulative selection: nature gives successive variations; man adds them up in certain directions useful to him. In this sense he may be said to make for himself useful breeds" (*Origin*, 30). Not just pigeons either: all kinds of animals and plants. And note how the selection is for features desired by the breeders. They are of a kind that is useful or admired.

> Youatt, who was probably better acquainted with the works of agriculturalists than almost any other individual, and who was himself a very good judge of an animal, speaks of the principle of selection as "that which enables the agriculturist, not only to modify the character of his flock, but to change it altogether. It is the magician's wand, by means of which he may summon into life whatever form and mould he pleases." Lord Somerville, speaking of what breeders have done for sheep, says:—"It would seem as if they had chalked out upon a wall a form perfect in itself, and then had given it existence." That most skilful breeder, Sir John Sebright, used to say, with respect to pigeons, that "he would produce any given feather in three years, but it would take him six years to obtain head and beak." (*Origin*, 30)

## Natural Selection

Human-caused change having been introduced, including what Darwin called "unconscious selection," where change occurs without the intent of the breeder – in other words, change without a mind directing the change – he then moved on (in Chapter Two) to the natural world. This took him into the second part of his overall argument, first establishing that in populations of organisms there is always a deal of variation – bigger, smaller, faster, slower, sturdier, more fragile.

> No one supposes that all the individuals of the same species are cast in the very same mould. These individual differences are highly important for us, as they afford materials for natural selection to accumulate, in the same manner as man can accumulate in any given direction individual differences in his domesticated productions. (*Origin*, 45)

Darwin, like others, had little knowledge of the sources of such variation, but an eight-year study of barnacles had convinced him that it always exists. Important was his conviction that, however caused, such variation never seems to satisfy needs. In this sense it is "random." Second, preparing the way for evolution, Darwin argued that natural populations differ, but how much they differ is in a sense arbitrary. The difference could be small, it could be big. Indeed, it is often unclear where a variety ends and a species begins. "I look at the term species, as one arbitrarily given for the sake of convenience to a set of individuals closely resembling each other, and that it does not essentially differ from the term variety, which is given to less distinct and more fluctuating forms" (*Origin*, 52). In other words, no need to assume major jumps – "saltations." Causes could and probably do bring on gradual change.

Now, in Chapters Three and Four, Darwin was ready for the heavy-duty arguments that lead to the causal forces of change. First, the "Struggle for Existence." At the end of the eighteenth century, there was already a major population explosion, which only intensified as the nineteenth century got underway. London, for instance, grew from about three-quarters of a million inhabitants in the middle of the eighteenth century to about a million and a half inhabitants by 1815. By 1860, the population was over three million. This growth in major part was brought on by the move to industrialism, where children (because they could work the machines) came early and frequently, as opposed to rural life where children were often postponed until family farms could be inherited.

However, the Anglican parson Thomas Robert Malthus was able to reassure people that this could not go on indefinitely. Population numbers increase geometrically – 1, 2, 4, 8, 16... – whereas food and space increase only arithmetically – 1, 2, 3, 4, 5.... There is going to be a crunch, and only some can survive – or, more significantly, contribute offspring to the next generation. The produce of the Earth

> ... may increase for ever and be greater than any assignable quantity; yet still the power of population being in every period so much superior, the increase of the human species can only be kept down to the level of the means of subsistence by the constant operation of the strong law of necessity, acting as a check upon the greater power. (*Essay*, 1826)

Darwin incorporated this argument in its entirety into his theory.

> A struggle for existence inevitably follows from the high rate at which all organic beings tend to increase. Every being, which during its natural lifetime produces several eggs or seeds, must suffer destruction during some period of its life, and during some season or occasional year, otherwise, on the principle of geometrical increase, its numbers would quickly become so inordinately great that no country could support the product. Hence, as more individuals are produced than can possibly survive, there must in every case be a struggle for existence, either one individual with another of the same species, or with the individuals of distinct species, or with the physical conditions of life. (*Origin*, 63)

Darwin used this argument as the springboard for a cause that produces indefinite change. Only a few getting through to the next generation is a natural equivalent of the breeder's method of producing change.

> HOW will the struggle for existence, discussed too briefly in the last chapter, act in regard to variation? Can the principle of selection, which we have seen is so potent in the hands of man, apply in nature? I think we shall see that it can act most effectually. Let it be borne in mind in what an endless number of strange peculiarities our domestic productions, and, in a lesser degree, those under nature, vary; and how strong the hereditary tendency is. Under domestication, it may be truly said that the whole organisation becomes in some degree plastic. Let it be borne in mind how infinitely complex and close-fitting are the mutual relations of all organic beings to each other and to their physical conditions of life. Can it, then, be thought improbable, seeing that variations useful to man have undoubtedly occurred, that other variations useful in some way to each being in the

great and complex battle of life, should sometimes occur in the course of thousands of generations? If such do occur, can we doubt (remembering that many more individuals are born than can possibly survive) that individuals having any advantage, however slight, over others, would have the best chance of surviving and of procreating their kind? On the other hand, we may feel sure that any variation in the least degree injurious would be rigidly destroyed. This preservation of favourable variations and the rejection of injurious variations, I call Natural Selection. (*Origin*, 80–81)

An important point here and a codicil. The point is that – something we shall later see caused much angst – although individual variations may appear randomly, selection-driven change is not random. It is in the direction of "useful" features or characteristics. Remember: "It would seem as if they had chalked out upon a wall a form perfect in itself, and then had given it existence." For the breeder, the features are those he or she wants – shaggier sheep or more melodious songbirds. In nature, the features are those that the possessor needs to outlive (out-reproduce) competitors: "adaptations."

We see these beautiful co-adaptations most plainly in the woodpecker and missletoe; and only a little less plainly in the humblest parasite which clings to the hairs of a quadruped or feathers of a bird; in the structure of the beetle which dives through the water; in the plumed seed which is wafted by the gentlest breeze; in short, we see beautiful adaptations everywhere and in every part of the organic world. (*Origin*, 60–1)

The codicil is that there is a secondary form of selection: sexual selection. Guided in his thinking by the fact that breeders select, on the one hand, for useful features like porkier pigs, and, on the other hand, for attractive features like more beautiful tail feathers, Darwin supposed that there is natural selection, producing adaptations for living, and sexual selection, producing adaptations useful towards the end of mate-attraction. "This depends, not on a struggle for existence, but on a struggle between the males for possession of the females; the result is not death to the unsuccessful competitor, but few or no offspring." He continues: "in many cases, victory will depend not on general vigour, but on having special weapons, confined to the male sex. A hornless stag or spurless cock would have a poor chance of leaving offspring" (*Origin*, 88).

After a couple of very brief hypothetical examples of how natural selection might be expected to work – Darwin instanced wolves chasing prey, deer, who would be under selective pressure to be slim, no excess weight to carry, and fast, able to outrun their victims – he rushed along to the implications of the second part of his argument. Here Darwin relied on what the eighteenth-century Scottish political economist Adam Smith in 1776 had called a "division of labour." Things will work far more efficiently if everyone does their own specialized job rather than attempting to be a jack of all trades. In nature, if organisms specialize, they will do better than otherwise: "the more diversified the descendants from any one species become in structure, constitution, and habits, by so much will they be better enabled to seize on many and widely diversified places in the polity of nature, and so be enabled to increase in numbers" (Darwin 1859, 112). This will lead eventually to many different species, all using their specific adaptations to succeed in the niches that they inhabit. And so, ultimately, we will get evolution. The fact of evolution. Using a metaphor that was there from the moment that Darwin became an evolutionist, we get a "tree of life" (Fig. 1.3 shows a sketch Darwin drew in a notebook, early in 1838).

> The affinities of all the beings of the same class have sometimes been represented by a great tree. I believe this simile largely speaks the truth.

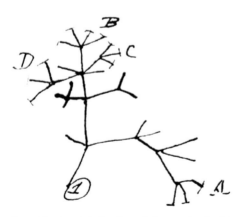

**Figure 1.3**  Darwin's sketch, made in a notebook, of the tree of life, showing that he had now become an evolutionist.

The green and budding twigs may represent existing species; and those produced during each former year may represent the long succession of extinct species. At each period of growth all the growing twigs have tried to branch out on all sides, and to overtop and kill the surrounding twigs and branches, in the same manner as species and groups of species have tried to overmaster other species in the great battle for life. (*Origin*, 129)

He continues:

As buds give rise by growth to fresh buds, and these, if vigorous, branch out and overtop on all sides many a feebler branch, so by generation I believe it has been with the great Tree of Life, which fills with its dead and broken branches the crust of the earth, and covers the surface with its ever branching and beautiful ramifications. (*Origin*, 130; Fig. 1.4)

The second part of the *Origin* now completed, Darwin had a couple of linking, albeit somewhat desultory, discussions (Chapters Five and Six), where he talked a little about the nature and possible causes of variation and looked at some

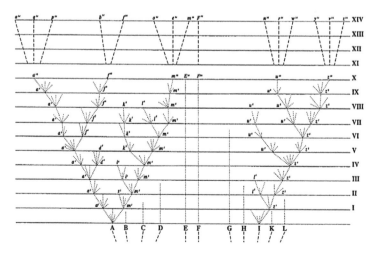

**Figure 1.4**    The tree of life (the only picture) in the *Origin*. Note that the point of the picture is to illustrate the division of labor, and how branching is ubiquitous.

possible objections towards what he had thus far argued. What is of interest (and subsequent importance) is a clarification at the end of Chapter Six. Darwin stressed that he did not think that all aspects of an organism necessarily must be adaptive, and he drew attention to something known to Aristotle, namely that there are isomorphisms (structural similarities; what Darwin's contemporary, the anatomist Richard Owen, in 1843 had called "homologies") between organisms of different kinds (Fig. 1.5). Using the terms "unity of type" (homology) and "conditions of existence" (adaptation), Darwin wrote:

> It is generally acknowledged that all organic beings have been formed on two great laws—Unity of Type, and the Conditions of Existence. By unity of type is meant that fundamental agreement in structure, which we see in organic beings of the same class, and which is quite independent of their habits of life. On my theory, unity of type is explained by unity of descent. The expression of conditions of existence, so often insisted on by the illustrious Cuvier, is fully embraced by the principle of natural selection. For natural selection acts by either now adapting the varying parts of each being to its organic and inorganic conditions of life; or by having adapted

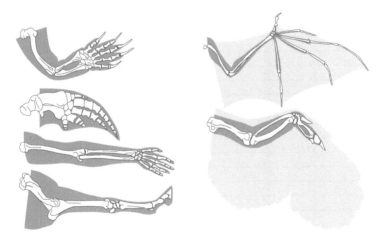

**Figure 1.5** Homology: isomorphisms between the bones and their ordering in different species.

them during long-past periods of time: the adaptations being aided in some cases by use and disuse, being slightly affected by the direct action of the external conditions of life, and being in all cases subjected to the several laws of growth. Hence, in fact, the law of the Conditions of Existence is the higher law; as it includes, through the inheritance of former adaptations, that of Unity of Type. (*Origin*, 206)

The importance of this passage becomes clearer as the *Origin* proceeds. On the one hand, the primacy of natural selection, and consequent adaptation, is affirmed; on the other hand, the isomorphisms between organisms widely different are acknowledged. This latter is important because, as Darwin knew full well from his work on barnacles, such isomorphisms or homologies are the key to ferreting out distant relationships and lines of descent: "phylogenies."

We now move (across more than half the length of the *Origin*) to the third part of Darwin's long argument. Darwin surveyed the various areas of the life sciences, showing how evolution through natural selection could explain much, and how, also, the explanations reflected back up to the causal mechanism of selection, confirming its existence and power. He started (his Chapter Seven) with behavior: "Instinct." Actually, it was not just behavior that interested Darwin but *social* behavior in particular. How do insects like the hymenoptera (ants, bees, wasps) function so harmoniously together? One thing that attracted his attention was the existence of sterile workers. He argued that, analogously to the farm, where sterile animals like the (castrated) ox can be improved by going back to the parent stock and breeding from those that have desirable offspring, so in the nest those fertile insects that have more and more efficient sterile offspring will be naturally selected.

This difficulty, though appearing insuperable, is lessened, or, as I believe, disappears, when it is remembered that selection may be applied to the family, as well as to the individual, and may thus gain the desired end. Thus, a well-flavoured vegetable is cooked, and the individual is destroyed; but the horticulturist sows seeds of the same stock, and confidently expects to get nearly the same variety; breeders of cattle wish the flesh and fat to be well marbled together; the animal has been slaughtered, but the breeder goes with confidence to the same family. (*Origin*, 237–8)

Darwin continued:

> Thus I believe it has been with social insects: a slight modification of structure, or instinct, correlated with the sterile condition of certain members of the community, has been advantageous to the community: consequently the fertile males and females of the same community flourished, and transmitted to their fertile offspring a tendency to produce sterile members having the same modification.

Not just sterile workers, but sterile workers of different castes:

> I believe that natural selection, by acting on the fertile parents, could form a species which should regularly produce neuters, either all of large size with one form of jaw, or all of small size with jaws having a widely different structure; or lastly, and this is our climax of difficulty, one set of workers of one size and structure, and simultaneously another set of workers of a different size and structure;—a graduated series having been first formed, as in the case of the driver ant, and then the extreme forms, from being the most useful to the community, having been produced in greater and greater numbers through the natural selection of the parents which generated them; until none with an intermediate structure were produced. (*Origin*, 241)

We move next (Chapter Eight) to "Hybridism." The main aim here was to show that, when it comes to questions of sterility, as one might expect from an evolutionary perspective (and primed by the discussion in Chapter Two), varieties slide in a more-or-less-gradated way into species – hybrids fertile, to hybrids less-than-fertile, to hybrids sterile. Natural selection does not play a big role in this discussion, but its importance picks up again as we turn to Chapters Nine and Ten on geology, and the nature and significance of fossils (Fig. 1.6 shows the fossil record as known at the time of the *Origin*, from Richard Owen's 1860 *Palaeontology*). What did not worry Darwin was the needed time for evolutionary change. In his mind, that battle was over. "It is hardly possible for me even to recall to the reader, who may not be a practical geologist, the facts leading the mind feebly to comprehend the lapse of time" (*Origin*, 282). We are talking millions – probably hundreds of millions – of years here.

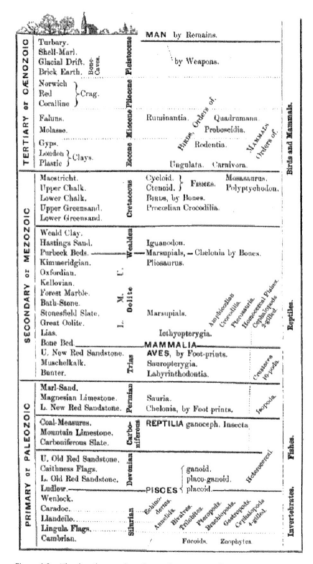

**Figure 1.6**    The fossil record as drawn by Owen in his *Palaeontology* (1860).

This given, there are still worries. Significantly, the first part of the chapter finds Darwin on his back foot, defending his theory against the charge that, because we do not find a smooth gradation of forms as we might expect if evolution through natural selection is the chief causal factor, it cannot be true. "Why then is not every geological formation and every stratum full of such intermediate links? Geology assuredly does not reveal any such finely graduated organic chain; and this, perhaps, is the most obvious and gravest objection which can be urged against my theory. The explanation lies, as I believe, in the extreme imperfection of the geological record" (*Origin*, 280). Ever inventive, ever persuasive, Darwin gave explanations backing this claim. Even he, though, admitted that some problems could not, at the moment, be solved. Most worrisome of all was that fairly sophisticated organisms just appeared in the fossil record, at the lowest levels, meaning the earliest times. Before that, nothing:

> ... if my theory be true, it is indisputable that before the lowest Silurian stratum was deposited, long periods elapsed, as long as, or probably far longer than, the whole interval from the Silurian age to the present day; and that during these vast, yet quite unknown, periods of time, the world swarmed with living creatures.
>
> To the question why we do not find records of these vast primordial periods, I can give no satisfactory answer. (*Origin*, 307)

Not all is gloom and doom, starting with the fact that the fossil record as it was suggested evolution. "New species have appeared very slowly, one after another, both on the land and in the waters." What was eye-catching was the way that as we go back in time, older organisms seemed to link up forms very different today – which is as we would expect given evolution. "It is a common belief that the more ancient a form is, by so much the more it tends to connect by some of its characters groups now widely separated from each other" (*Origin*, 330). Of course, this is not universally true. "Yet if we compare the older Reptiles and Batrachians, the older Fish, the older Cephalopods, and the eocene Mammals, with the more recent members of the same classes, we must admit that there is some truth in the remark" (*Origin*, 330–1). What is exciting is the claim "that ancient animals resemble to a certain extent the embryos of recent animals of the same classes; or that the geological succession of extinct forms is in some degree parallel to the embryological development of recent forms." Darwin claimed not to be entirely convinced of the truth of this claim: "Yet I fully expect to see it hereafter confirmed, at least in regard to subordinate groups, which have branched off from each other within comparatively recent times" (*Origin*, 338). He added that he would have more to say in a later chapter (Chapter Thirteen), where he would spell out how adaptations might be added at a later stage of individual development (ontogeny; as opposed to phylogeny, group development), and so the early forms would remain the same, even as newer forms retained the early forms while embryos and then went on to differ as the adults.

Starting now to warm to his themes, in Chapters Eleven and Twelve Darwin took on the facts of geographical distribution. He was not keen on the idea that different species were created from scratch in situ around the globe. He thought this highly improbable. "Nevertheless the simplicity of the view that each species was first produced within a single region captivates the mind. He who rejects it, rejects the *vera causa* of ordinary generation with subsequent migration, and calls in the agency of a miracle" (*Origin*, 352). Much space was devoted, again, to arguing against problems, here particularly how organisms could have moved naturally over long distances, often including travel over vast oceans. Then comes the positive case: ocean islands.

> The most striking and important fact for us in regard to the inhabitants of islands, is their affinity to those of the nearest mainland, without being actually the same species. Numerous instances could be given of this fact. I will give only one, that of the Galapagos Archipelago, situated under the equator, between 500 and 600 miles from the shores of South America. Here almost every product of the land and water bears the unmistakeable stamp of the American continent. (*Origin*, 397–8)

How could this possibly be, save through evolution?

> Why should this be so? why should the species which are supposed to have been created in the Galapagos Archipelago, and nowhere else, bear so plain a stamp of affinity to those created in America? There is nothing in the conditions of life, in the geological nature of the islands, in their height or climate, or in the proportions in which the several classes are associated together, which resembles closely the conditions of the South

American coast: in fact there is a considerable dissimilarity in all these respects. (*Origin*, 398)

The clincher, if one be needed, is that the denizens of the islands are similar, but different. There were original invaders, who then moved from island to island and diversified.

Moving along now rapidly towards the conclusion, Chapter Thirteen covers classification, morphology, embryology, and rudimentary organs. Classification, based on the Linnaean system, links organisms in ever broader groups – the organism is a member of a species, which then is grouped with species with similar organisms into genera (the plural of "genus"), and so all the way up to kingdoms. The reason? Evolution through natural selection. In truth, "natural selection, which results from the struggle for existence, and which almost inevitably induces extinction and divergence of character in the many descendants from one dominant parent-species, explains that great and universal feature in the affinities of all organic beings, namely, their subordination in group under group" (*Origin*, 433). Morphology, the anatomical nature of organisms, is (as we anticipated earlier) likewise explained by Darwinian theory. Most significant is what we have seen Owen calling "homologies," the isomorphisms between different animals. There is no point to them. They serve no function. Using language we shall explore later, they have no "final causes."

Nothing can be more hopeless than to attempt to explain this similarity of pattern in members of the same class, by utility or by the doctrine of final causes. The hopelessness of the attempt has been expressly admitted by Owen in his most interesting work on the 'Nature of Limbs.' On the ordinary view of the independent creation of each being, we can only say that so it is;—that it has so pleased the Creator to construct each animal and plant. (*Origin*, 435)

So why the similarities? What is the significance of homology? "The explanation is manifest on the theory of the natural selection of successive slight modifications—each modification being profitable in some way to the modified form, but often affecting by correlation of growth other parts of the organisation. In changes of this nature, there will be little or no tendency to modify the original pattern, or to transpose parts." Reverting again to Owen's view of things, Darwin says: "If we suppose that the ancient progenitor, the archetype as it may be called, of all mammals, had its limbs constructed on the existing general pattern, for whatever purpose they served, we can at once perceive the plain signification of the homologous construction of the limbs throughout the whole class" (*Origin*, 435).

Figure 1.7 shows Owen's archetype.

Embryology was a favorite of Darwin. Why is it that, picking up on a question left dangling earlier, very different adult organisms so often have very similar embryos? It is simply because the embryos in the egg or the womb are not much affected by natural selection responding to changed circumstances, but as the embryos grow, natural selection gets to work. Moreover: "As the embryonic state of each species and group of species partially shows us the structure of their less modified ancient progenitors, we can clearly see why ancient and extinct forms of life should resemble the embryos of their descendants,—our existing species" (*Origin*, 449).

And so, after a quick look at rudimentary organs – the flotsam and jetsam of the evolutionary process – we are ready for Chapter Fourteen: "Recapitulation and Conclusion." Missing thus far in the argumentation of the *Origin* has been our own species. It remains missing except for a glancing reference. "Light will be thrown on the origin of man and his history" (488). And then the climax – the most famous passage in the whole of science.

> It is interesting to contemplate an entangled bank, clothed with many plants of many kinds, with birds singing on the bushes, with various

**Figure 1.7**   Owen's vertebrate archetype. This is a kind of ideal, Platonic Form of the skeleton of a vertebrate. Then, through the action of natural selection, the vertebrate is modified in various ways, giving rise to homologies.

insects flitting about, and with worms crawling through the damp earth, and to reflect that these elaborately constructed forms, so different from each other, and dependent on each other in so complex a manner, have all been produced by laws acting around us. These laws, taken in the largest sense, being Growth with Reproduction; Inheritance which is almost implied by reproduction; Variability from the indirect and direct action of the external conditions of life, and from use and disuse; a Ratio of Increase so high as to lead to a Struggle for Life, and as a consequence to Natural Selection, entailing Divergence of Character and the Extinction of less-improved forms. Thus, from the war of nature, from famine and death, the most exalted object which we are capable of conceiving, namely, the production of the higher animals, directly follows. There is grandeur in this view of life, with its several powers, having been originally breathed into a few forms or into one; and that, whilst this planet has gone cycling on according to the fixed law of gravity, from so simple a beginning endless forms most beautiful and most wonderful have been, and are being, evolved. (*Origin*, 489–90)

# 2 Organicism and Mechanism: Rival Root Metaphors

Natural selection. I am an evolutionist, which means that, to understand the present, we must dig into the past. That holds for culture as much as for biology. So, taking my own advice, where do we end up? Or, more precisely, where do we start off? As always, when dealing with Western culture, we begin with the Greeks, Plato and Aristotle. Neither of them was an evolutionist. Indeed, rather like the Buddhists, they believed that the (physical) world is eternal: no beginning, no end. But they did have much to say of great interest to our inquiry.

## Organicism

Ask about the Greeks' world picture. How did they see things? By this I mean, how did they organize their thinking about the physical world? Today, it is practically a commonplace that the key to such organization is metaphor. Our understanding is not (as TV's *Dragnet* would demand) "the facts, just the facts." We put the information into a picture, a conceptual scheme. If we are having an argument, then typically we think of it as a war: I went right at him. Offensive. I pulled back to reconsider. Retreat and regroup. I tried a different line of attack. New weapons. We agreed to disagree. Armistice. When we put it all together, then we have what the linguist Stephen Pepper in 1942 called "root metaphors." "A man desiring to understand the world looks about for a clue to its comprehension. He pitches upon some area of commonsense fact and tries to understand other areas in terms of this one. The original area becomes his basic analogy or root metaphor." These are much akin to what Thomas Kuhn called "paradigms."

What was the root metaphor of the Greeks? Very naturally, it was an organism. Everyone was living in a world close to nature: birth in the spring; adulthood in the summer; growing old in the fall; coming to an end and death in the winter. In his dialogue, the *Timaeus*, Plato is explicit.

> Why did the Creator make the world? . . . He was good, and therefore not jealous, and being free from jealousy he desired that all things should be like himself. Wherefore he set in order the visible world, which he found in disorder. Now he who is the best could only create the fairest; and reflecting that of visible things the intelligent is superior to the unintelligent, he put intelligence in soul and soul in body, and framed the universe to be the best and fairest work in the order of nature, and the world became a living soul through the providence of God.
>
> In the likeness of what animal was the world made? . . . The form of the perfect animal was a whole, and contained all intelligible beings, and the visible animal, made after the pattern of this, included all visible creatures. (*Plato Complete Works*, 1236)

"Creator" is not quite the right word. The force behind all of this did not create; it designed. In his dialogue the *Phaedo*, about Socrates's last day on Earth, Plato makes it clear that to understand living things – including now the physical world – we must think in terms of functions, of ends. The hand and the eye grow, but why do they exist at all?

> One day I heard someone reading, as he said, from a book of Anaxagoras, and saying that it is Mind that directs and is the cause of everything. I was delighted with this cause and it seemed to me to be good, in a way, that Mind should be the cause of all. I thought that if this were so, the directing Mind would direct everything and arrange each thing in the way that was best. (*Plato Complete Works*, 84)

Whether in one fell swoop in time or, more likely, as a kind of principle of ordering, the Designer (which Plato called the "Demiurge") put things in place, having them function for the purpose of efficient life: the eye to see, the hand to grasp. Linking this up with his basic ontology, the Theory of Forms, Plato identified the Demiurge with the most important of them all, the Form of the Good, from which all else flows.

Aristotle's world view ruled out a designing force behind the world. His Supreme Being, the Unmoved Mover, towards which everything is directed, is a Perfect Being, and hence does the only thing that a Perfect Being can do, namely contemplate its own perfection. It has no knowledge of or interest in the physical world. So, there is no external designer, but internal design. The thinking about the world continues to be organismic, thinking in terms of purposes, of ends, of what Aristotle called "final causes," as well as in terms of causes that get things going, "efficient causes." "In dealing with respiration we must show that it takes place for such or such a final object; and we must also show that this and that part of the process is necessitated by this and that other stage of it" (*Complete Works of Aristotle*, 999).

Note the difference between Plato and Aristotle. Both Plato and Aristotle are thinking in terms of ends; in today's language, they are both thinking "teleologically" (that is, in terms of design and purpose). For Plato, the cause of this end-direction is a designing mind. The Demiurge wanted animals to be able to see, so He designed eyes for this end. For Aristotle, the motive force was a special kind of force, one that is end-directed. The force is part of nature, but it is not an efficient cause. In a sense, the Unmoved Mover is responsible, for it is that towards which all strive, but it is not a cause that makes and maintains.

It is easy to see how all this fit readily into a Christian context. For St Augustine, influenced by Platonism, it was obvious that the world was designed by an all-powerful Mind. For St Thomas Aquinas, an Aristotelian, it was no less obvious that there are forces in the world directed towards ends. For both of these great thinkers, everything, organic and inorganic, served ends. This brings us to the Scientific Revolution which, conveniently, we can date from the beginning of the sixteenth century and the arrival of Copernicus's heliocentric theory of the universe, to the end of the seventeenth century and Newton's causal explanation of everything in terms of his laws of motion and that of gravitational attraction.

## Mechanism

More than anything, the Scientific Revolution was a change of root metaphor, from that of an organism to that of a machine. This is no great surprise really, as one looks back and sees how machines – above all the clock – were starting to move into everyday life and show their importance for better living. As Thomas Kuhn pointed out, changes in paradigms – or metaphors as he later identified them with – do not just happen. On the one hand, the old metaphor starts to break down. On the other hand, there must be a new metaphor to replace the old. We can certainly see that, by the end of the fifteenth century, the old metaphor was breaking down. The mineralogist Georg Agricola, 1495–1555, put his finger on a major problem. First, he gave the traditional argument:

> The earth does not conceal and remove from our eyes those things which are useful and necessary to mankind, but on the contrary, like a beneficent and kindly mother she yields in large abundance from her bounty and brings into the light of day the herbs, vegetables, grains, and fruits, and the trees. The minerals on the other hand she buries far beneath in the depth of the ground; therefore, they should not be sought. But they are dug out by wicked men who, as the poets say, "are the products of the Iron Age." (De Re Metallica, 6–7; the translation is by the future president of the United States, Herbert Hoover, who was a mining engineer at the time.)

Then Agricola went after the argument. In his view, it is just silly to say that mining for metals is wrong because it upsets Mother Earth. That is taking metaphor way too far. "If there were no metals, men would pass a horrible and wretched existence in the midst of wild beasts; they would return to the acorns and fruits and berries of the forest."

This said, there was no overnight conversion. Johannes Kepler, one of the heroes of the Scientific Revolution, was an enthusiastic organicist, writing in 1619:

> as the body displays tears, mucus, and earwax, and also in places lymph from pustules on the face, so the Earth displays amber and bitumen; as the bladder pours out urine, so the mountains pour out rivers; as the body produces excrement of sulphurous odor and farts which can even be set on fire, so the Earth produces sulphur, subterranean fires, thunder, and lightning; and as blood is generated in the veins of an animate being, and with it sweat, which is thrust outside the body, so in the veins of the Earth are generated metals and fossils, and rainy vapor. (Harmony, 363–64)

However, increasingly the new root metaphor, that of a machine, found favor with empirical scientists, and became their preferred method of conceptualizing their work. And not just in physics: as the French philosopher René Descartes realized in 1637, William Harvey's claim that the heart is a pump is a mechanistic way to understand the human body. "This will hardly seem strange to those who know how many motions can be produced in automata or machines which can be made by human industry, although these automata employ very few wheels and other parts in comparison with the large number of bones, muscles, nerves, arteries, veins, and all the other component parts of each animal" (*Discourse on Method*, 41).

Robert Boyle, the chemist and natural philosopher, made the definitive statement in 1686.

> [The world] is like a rare clock, such as may be that at Strasbourg, where all things are so skillfully contrived that the engine being once set a-moving, all things proceed according to the artificer's first design, and the motions of the little statues that at such hours perform these or those motions do not require (like those of puppets) the peculiar interposing of the artificer or any intelligent agent employed by him, but perform their functions on particular occasions by virtue of the general and primitive contrivance of the whole engine. (*Free Enquiry*, 12–13)

What we have is matter in motion, going through its cycles, governed by eternal, unbreakable laws. Not coming from anywhere. Not going anywhere. Without purpose. Without value. Of course, an individual machine has a purpose. A clock is for telling time. But understood as part of science, the machine has no purpose. At least it has no absolute (external) purpose. There may be internal purposes: for instance, the function of this cog is to drive this belt. But that is our judgment, not an absolute, found in nature. In the context of science, it is all just matter in motion. As said by the English philosopher of science Francis Bacon in 1605, final causes are like vestal virgins, beautiful but sterile. And God? In the words of Eduard Jan Dijksterhuis, one of the great historians of the Scientific Revolution, God has become a "retired engineer."

Does this mean that post-Revolution science, modern science, has become atheistic? Not necessarily. Copernicus was in minor orders. Descartes was a sincere Catholic; Newton a member of the Anglican Church. What it does mean, however, is that there is going to be somewhat of a push away from theism, belief in a God who intervenes (miraculously) in His creation, to deism, where God is an Unmoved Mover who created and set everything in motion and then sat back without intervening.

## The Problem of Organisms

Surely this does mean an end to Christianity, with its story of God's intervening in the world for our salvation? Not quite. There was still the problem of organisms. It is all very well to say that the moon has no purpose, just endlessly circulating around the Earth, held in place by gravitational attraction. Whereas before one might have said that the rain falls to water the crops, now the rain just falls. It does not fall to do anything. That the crops might benefit is another, different, matter. But Scientific Revolution or not, organisms are not like the moon or rain. The heart exists in order to pump the blood, because we need it to go on living. The final cause of the heart is blood pumping. The same goes for other features such as the hand and the eye. Say what you like, organisms still seem to be designed. You need absolute values to understand living things.

Boyle recognized the problem. His solution in 1688 was to kick that aspect of organisms out of science and into religion:

> For there are some things in nature so curiously contrived, and so exquisitely fitted for certain operations and uses, that it seems little less than blindness in him, that acknowledges, with the Cartesians [followers of Descartes], a most wise Author of things, not to conclude, that, though they may have been designed for other (and perhaps higher) uses, yet they were designed for this use." (*Disquisition*, 16)

Boyle continues: the supposition "that a man's eyes were made by chance, argues, that they need have no relation to a designing agent; and the use, that a man makes of them, may be either casual too, or at least may be an effect of his knowledge, not of nature's." This cannot be. Step forward, God. "It is rational, from the manifest fitness of some things to cosmical or animal ends or uses, to infer, that they were framed or ordained in reference there-unto by an intelligent and designing agent" (19).

It was a compromise that let people go on doing biological science. Some were quite happy with the state of affairs. In 1802 Archdeacon William Paley, writer of textbooks at the end of the eighteenth century, positively wallowed in the way that organisms demand a Designer.

> I know no better method of introducing so large a subject, than that of comparing a single thing with a single thing: an eye, for example, with a telescope. As far as the examination of the instrument goes, there is precisely the same proof that the eye was made for vision, as there is that the telescope was made for assisting it. They are made upon the same principles; both being adjusted to the laws by which the transmission and refraction of rays of light are regulated. (*Natural Theology*, 22)

Others, David Hume for instance, more reluctantly were drawn to acknowledge the likelihood that a Designer was involved. He wrote in 1779:

> That the works of Nature bear a great analogy to the productions of art, is evident; and according to all the rules of good reasoning, we ought to infer, if we argue at all concerning them, that their causes have a proportional analogy. But as there are also considerable differences, we have reason to suppose a proportional difference in the causes; and in particular, ought to attribute a much higher degree of power and energy to the supreme cause, than any we have ever observed in mankind. Here then the existence of a DEITY is plainly ascertained by reason: and if we make it a question, whether, on account of these analogies, we can properly call him a mind or intelligence, notwithstanding the vast difference which may reasonably be supposed between him and human minds; what is this but a mere verbal controversy? (*Dialogues*)

At the other end of Europe, Immanuel Kant was tormented by the problem. He was a firmly committed Newtonian. All the way, it is laws of nature, forever going through their motions, mindlessly. And yet, organisms are the spanner in the works. They do seem to show purpose. In the end, Kant decided that final-cause attributions are heuristic. They help us think about organisms. They are "regulative." They are not part of reality. They are not "constitutive." As he wrote in 1790:

> The concept of a thing as in itself a natural end is therefore not a constitutive concept of the understanding or of reason, but it can still

be a regulative concept for the reflecting power of judgment, for guiding research into objects of this kind and thinking over their highest ground in accordance with a remote analogy with our own causality in accordance with ends; not, of course, for the sake of knowledge of nature or of its original ground, but rather for the sake of the very same practical faculty of reason in us in analogy with which we consider the cause of that purposiveness. (*Critique of Judgement*, 36)

This is all very well. But it does mean that biology is forever condemned to be second-rate. "[W]e can boldly say that it would be absurd for humans even to make such an attempt or to hope that there may yet arise a Newton who could make comprehensible even the generation of a blade of grass according to natural laws that no intention has ordered; rather, we must absolutely deny this insight to human beings" (37). And this brings in Charles Darwin, because above all he wanted to be the Newton of the blade of grass, and he thought that natural selection was the key to achieving that end.

## Evolution

Let us introduce evolution into our story. A necessary condition for evolution is adequate time. By the mid-eighteenth century, that was becoming a probability. Geology and fossil discoveries started to push the beginning of life back a lot further than the traditional 6,000 years of Genesis. Almost paradoxically, Christianity played a major role in the move to evolution. As we have seen, the Greeks did not have a history with a beginning moment in time. The Jews, and subsequently the Christians, did have such a history, from nothing to the beings "made in the image of God." So, as people started to move towards a more deistic view of the godhead, it was natural to think that perhaps there could have been a developing process of origin, with humans as the apotheosis. Thinking this way, there was no need to start from scratch.

The idea of a scale of beings, from the simplest to the most complex, humans, then angels, and then God, had a long history (Fig. 2.1). It was just a matter of turning it from a staircase to an escalator. Here, the move from theism to deism paid major dividends. We humans were no longer seen as helpless, saved only by the sacrifice on the Cross, the ultimate Divine intervention, the miracle of

**Figure 2.1**  The Great Chain of Being, from Ramon Lull's *Ladder of Ascent and Descent of the Mind*: God, angels, heaven, humans, beasts, plants, flame, rocks.

miracles. Now, as people explored the world, as they moved towards industrialism, as they started to improve their own lives – housing, health, education – the
belief that we do not need God intervening, that we can do it all ourselves,
became more and more compelling. In other words, there was a move from
Providence to Progress. Evolution, the idea of organisms growing up from blobs
to humans, under their own steam as it were (a nice machine metaphor) was
really Progress written into the organic world, including into an increasingly well-
known fossil record.

If not strictly the first, Charles Darwin's grandfather, Erasmus Darwin, eminent
physician and poet, at the end of the eighteenth century became an enthusiast.
In 1803 he wrote:

> Organic Life beneath the shoreless waves
> Was born and nurs'd in Ocean's pearly caves;
> First forms minute, unseen by spheric glass,
> Move on the mud, or pierce the watery mass;
> These, as successive generations bloom,
> New powers acquire, and larger limbs assume;
> Whence countless groups of vegetation spring,
> And breathing realms of fin, and feet, and wing.
> Thus the tall Oak, the giant of the wood,
> Which bears Britannia's thunders on the flood;
> The Whale, unmeasured monster of the main,
> The lordly Lion, monarch of the plain,
> The Eagle soaring in the realms of air,
> Whose eye undazzled drinks the solar glare,
> Imperious man, who rules the bestial crowd,
> Of language, reason, and reflection proud,
> With brow erect who scorns this earthy sod,
> And styles himself the image of his God;
> Arose from rudiments of form and sense,
> An embryon point, or microscopic ens!
>
> (*The Temple of Nature*, 1, 11, 295–314)

Lest there be any doubt, Darwin explicitly drew the link between human
Progress and evolution. The latter "is analogous to the improving excellence

observable in every part of the creation; such as the progressive increase of the wisdom and happiness of its inhabitants," as he wrote about five years before the end of the eighteenth century (*Zoonomia*, II, 247–48).

## Natural Selection

The actual cause of evolution was a mystery. Erasmus Darwin opted for a version of Lamarckism, the inheritance of acquired characteristics. Basically, one gets the feeling that causes were not of prime importance to him – very much unlike his grandson, Charles Darwin. By the time Charles was born and was growing up, the Darwin family was very comfortable, thanks to Charles Darwin's mother being the daughter of Josiah Wedgwood, he of pottery fame and one of the most successful industrialists of his age (Browne 1995). Naturally, therefore, the young Charles was sent to an English public (read "private") school, and from there (via an unsuccessful attempt to study medicine at the University of Edinburgh) to the University of Cambridge (1828–31), with the intention that he would be ordained as a minister in the Anglican Church. That plan was quietly dropped when Charles Darwin got the opportunity to join the British warship HMS *Beagle*, as ship's naturalist, for five years (1831–36), as it mapped the coast of South America, ending the trip by going on around the world and visiting Australasia and South Africa, before returning home.

Even before he left Cambridge to join the *Beagle*, Darwin was an enthusiastic Newtonian, meaning that the world must be seen as a machine, working unceasingly, governed by eternal regularities. The greatest scientist of all time, Isaac Newton, was a Cambridge graduate. At Cambridge, Darwin's interests drew him towards the faculty who were scientists. They were, as might be expected, Newton promotors – "groupies" would not be too strong a term. Darwin's mentors, especially the botanist John Henslow and the historian and philosopher of science William Whewell (pronounced "Hule" as in "mule"), would have reiterated again and again the all-importance of explaining through law. Reinforcing this, just as he set out on his five-year trip, Darwin read a work (*Preliminary Discourse on the Study of Natural Philosophy*, published in 1830) by another ardent Newtonian, the astronomer John F. W. Herschel.

To top it off, a living scientist whom they all revered was the Scottish lawyer-turned-geologist, Charles Lyell, whose *Principles of Geology* (published in 1830–33) was just appearing. Darwin took the first volume with him, and the succeeding two were shipped out to him in South America. Lyell argued that unbroken law, working in near-infinite stretches of time, was enough – through rainfall, earthquakes, volcanoes and more – to produce all that we see around us: mountains, plains, valleys, streams, lakes, oceans. Lyell was an ardent deist – no miracles – and, away from home, with no one there to keep him on the right theological track, this clearly had an effect on the young Darwin. He moved to deism, a theology that stayed with him right through the writing of the *Origin* and, under the influence of his ardent supporter Thomas Henry Huxley, faded to agnosticism only in the last decade of his life.

Returning to England in 1836, it takes nothing from Darwin's genius to say that the route to discovery, first evolution and then natural selection, was straightforward and irresistible. On the one hand, it was a major prize hanging in front of a very ambitious young scientist. For all their pushing law, when it came to organisms and their design-like nature, everyone backed off from origins. In 1837, Whewell played a Boyle-like trick of pushing it all into religion – science is silent and "points upwards." Lyell played an even older trick. He said nothing! On the other hand, as a teenager, Darwin had read his grandfather's evolutionary work. He was primed to search for and find a naturalistic solution to the origins question. On cataloguing his avian specimens from the Galapagos Archipelago (in the Pacific and visited on the way home), Darwin realized that the birds on the islands were similar to those on the mainland, but slightly different. The same was true of birds on different islands – similar but slightly different. The obvious answer is that the birds came from the mainland to the archipelago, then they scattered from island to island, all the while diverging. They evolved.

For Darwin, as a Newtonian, this now meant that the hunt was on for causes. But not any old cause; a cause that could explain the design-like features of organisms. As a proto-parson at Cambridge, Darwin had read his Paley. As he wrote in a private notebook: "We never may be able to trace the steps by which the organization of the eye passed from simpler stages to more perfect preserving its relations. the wonderful power of adaptation given to organization. – This really perhaps greatest difficulty to whole theory" (Charles *Darwin's Notebooks*, notebook C, 175). We saw in the last chapter that Darwin came from rural England, and it was not long before he saw that the key to change lies in selective breeding. He even read a short pamphlet by Sebright that drew an explicit connection between such breeding and the possibility of a likewise phenomenon in nature:

> A severe winter, or a scarcity of food, by destroying the weak or unhealthy, has all the good effects of the most skilful selection. In cold and barren countries no animal can live to the age of maturity, but those who have strong constitutions; the weak and the unhealthy do not live to propagate their infirmities, as is too often the case with our domestic animals. To this I attribute the peculiar hardiness of the horses, cattle, and sheep, bred in mountainous countries, more than their having been inured to the severity of climate. . . (*Art of Improving*, 15–16)

This was written in 1809 by Sir John Sebright, to whom Darwin refers in the first chapter of the *Origin*. Darwin read the passage and, in the margin, marked its importance:

> Sir J. Sebright – pamphlet most important showing effects of peculiarities being long in blood. ++ thinks difficulty in crossing race – bad effects of incestuous intercourse. – excellent observations of sickly offspring being cut off so that not propagated by nature. – Whole art of making varieties may be inferred from facts stated. — (*Charles Darwin's Notebooks*, C, 133)

There is no implication here of plagiarism. In the pamphlet, Sebright gives no indication that he sees any profound or wide-ranging evolutionary implications.

But how could one get selection to work, on a regular basis, in nature? At the end of September 1838, Darwin read Malthus on population. At once he had his answer. Again, from a private notebook:

> Population is increase at geometrical ratio in <u>far shorter</u> time than 25 years – yet until the one sentence of Malthus no one clearly perceived the great check amongst men. – there is spring, like food used for other purposes as wheat for making brandy. – Even <u>a few</u> years plenty, makes population in Men increase & an <u>ordinary</u> crop causes a dearth. take

> Europe on an average every species must have same number killed year with year by hawks, by cold &c. – even one species of hawk decreasing in number must affect instantaneously all the rest. – The final cause of all this wedging, must be to sort out proper structure, & adapt it to changes. – to do that for form, which Malthus shows is the final effect (by means however of volition) of this populousness on the energy of man. One may say there is a force like a hundred thousand wedges trying force ~~into~~ every kind of adapted structure into the gaps ~~of~~ in the oeconomy of nature, or rather forming gaps by thrusting out weaker ones. (*Charles Darwin's Notebooks*, D 135e; underlining and crossings out are Darwin's)

Note the amusing paradox here. Darwin may have made the move from theism to deism, and this was crucial in his route to discovery; but, without Christian theism, he would have been nowhere. The very search for origins; the recognition of the design-like features of organisms; and now the influence of Malthus (an ordained Anglican), who put his argument to the struggle in a Christian theological context. It is God's way of getting us to work. Without the need of effort, we could spend our whole lives as philosophy graduate students. Then, moving on, we have the division of labor. It is God who sees that selfishness acts for the good of the group. And then finally, of course, there is the tree of life – theism incorporated into the theory of a Newtonian deist.

It was not long before Darwin was speculating about selection and its effects. Showing that from the first Darwin never doubted that his theory would apply to us humans, the first explicit reference is to our mental abilities.

> An habitual action must some way affect the brain in a manner which can be transmitted. – this is analogous to a blacksmith having children with strong arms. – The other principle of those children, which chance? produced with strong arms, outliving the weaker ones, may be applicable to the formation of instincts, independently of habits. (*Charles Darwin's Notebooks*, N 42)

Note that Darwin was, and ever was, a Lamarckian. He just thought it could never be cause enough on its own. A comment, before the discovery of selection, when he had only Lamarckism to rely on: "Wax of Ear, bitter perhaps to prevent insects lodging there, now these exquisite adaptations can hardly be accounted for by my method of breeding there must be some cor[r]elation, but the whole mechanism is so beautiful" (*Charles Darwin's Notebooks*, C 174).

## Completing the Scientific Revolution

Pause for a moment and ask what it was that Darwin had intended. Think again about Kant's worries and why he denied there could ever be a "Newton of the blade of grass." Organisms demand a teleological understanding. We encountered that notion in the last chapter, and you will have noticed it again just above, when, in appreciating the force of selection, Darwin wrote of the end result of the Malthusian pressure. Final causes. And this is exactly what the machine metaphor excludes. The best one can do is to follow Boyle and kick the problem out of science and into religion; or follow Kant and argue that final-cause talk is not really part of the science but just a heuristic for worried biologists. Hence, Darwin's task was not merely to find a cause of evolution but to show that it could explain final causes within the machine metaphor. One way would have been to cheat and simply refuse to answer final-cause questions. The heart pumps, but don't ask why. As a student of Paley, this was simply not an option for Darwin. Final causes were real, and, in the *Origin*, without embarrassment, he made use of them.

Take the fact that the cuckoo lays its eggs in the nests of other birds.

> It is now commonly admitted that the more immediate and *final cause* of the cuckoo's instinct is, that she lays her eggs, not daily, but at intervals of two or three days; so that, if she were to make her own nest and sit on her own eggs, those first laid would have to be left for some time unincubated, or there would be eggs and young birds of different ages in the same nest. If this were the case, the process of laying and hatching might be inconveniently long, more especially as she has to migrate at a very early period; and the first hatched young would probably have to be fed by the male alone ... Now let us suppose that the ancient progenitor of our European cuckoo had the habits of the American cuckoo; but that occasionally she laid an egg in another bird's nest. If the old bird profited by this occasional habit, or if the young were made more vigorous by advantage having been taken of the mistaken maternal instinct of another bird, than by their own mother's care, encumbered as she can hardly fail to be by having eggs and young of different ages at the same time; then the old birds or the fostered young would gain an advantage. And analogy would lead me to believe, that the young thus reared would be apt to

follow by inheritance the occasional and aberrant habit of their mother, and in their turn would be apt to lay their eggs in other birds' nests, and thus be successful in rearing their young. By a continued process of this nature, I believe that the strange instinct of our cuckoo could be, and has been, generated. (*Origin*, 216–18, my italics)

Take apart what is happening here. Some mother cuckoos lay their eggs (presumably at first by chance) in the nests of other birds. They reproduce more effectively – they have more offspring – than those mother cuckoos who raise their own young. In a while, this habit becomes innate, and the successful have completely eliminated the unsuccessful. Every mother cuckoo, by instinct, lays her eggs in the nests of other birds. The "final cause" of this instinct is more successful egg laying. Notice, however, that although we are referring to future events – mother cuckoo lays her eggs now, and down the road she has more viable offspring – we are not appealing to future causes. The causes are all in the past or the present. In the past, the successful mother cuckoo had ancestors who laid their eggs in the nests of others, and now she does the same. Obviously, we could be mistaken in assuming that the future will be like the past. An environmental change, for instance, means that birds in alien nests will never survive. The supposed final cause no longer works as a final cause. In the case of humans, for instance, the final cause of a sweet tooth is that those of our Pleistocene ancestors who searched out bees' nests, and got the honey, out-reproduced those that did not. Today, however, a sweet tooth leads to obesity (given the ready access to sugar), and that is disastrous. The point is that usually things do stay more or less the same, and so we can get away with using the future to make sense of past actions.

This is teleology without tears. No need of an external designer, or External Designer. No need of an inner force pushing us into a desirable future. It is all a question of cause and effect. Sexual intercourse (cause) produces humans (effect), who then grow and in turn act as cause to bring on sex (effect but also original cause), which then produces humans (effect), and so it has gone on for millions of years. The cause brings on the effect, which then acts as a cause to bring on the original cause, which again brings on the effect. We conceptualize it as the cause existing to bring on the future effect, because (without disturbance) the cause does bring on the effect, which means that now the effect can bring on the originating cause. This is something, incidentally, that Kant recognized: "An organized product of nature is that in which everything is an end and reciprocally a means as well. Nothing in it is in vain, purposeless, or to be ascribed to a blind mechanism of nature." What Kant could not see is that this gives him what he wants, a blindness presumably at least in part because he was not an evolutionist. He could not accept evolution because he could not see his way around final cause. Conversely, he could not see his way around final cause because he was not an evolutionist. Darwin was an evolutionist, and, with his causal process of natural selection, he saw that he could incorporate final cause within a machine-metaphor picture of the world. Darwin completed the Scientific Revolution.

# 3 "The Non-Darwinian Revolution?"

Among the many books authored by Peter Bowler, the eminent historian of evolutionary biology, three stand out: *The Eclipse of Darwinism*; *The Non-Darwinian Revolution: Reinterpreting a Historical Myth*; and *Darwin Deleted: Imagining a World without Darwin*. Bluntly, he says: "there is now a substantial body of literature to convince anyone that the part of Darwin's theory now recognized as important by biologists had comparatively little impact on late nineteenth century thought". I cannot say Bowler is entirely wrong. Indeed, in *The Darwinian Revolution: Science Red in Tooth and Claw*, I contributed to this "body of literature," and my book was quite openly a synthesis of the state of Darwinian play in the second half of the twentieth century. But is this the end of the story, and if it is, why is it the end of the story? Today, as Bowler also recognizes, we accept the finding of natural selection as a major scientific achievement, up there with relativity theory. Let us pick up on this paradox.

## Inadequacy as Science

The Bowler thesis, as we might call it, has limited scope. It is directed at Darwinian natural selection. Everyone agrees that the *Origin* was a – if not the – major factor in the spread of evolutionary ideas, in Western countries, in the second half of the nineteenth century. The skepticism is about the cause, natural selection – a skepticism stemming from the fact that, from the Victorian perspective, it was going to be inadequate at the level of science. It simply could not do what it claimed to be able to do – or, if it could, it was not needed, a handicap indeed to understanding. No one was going to deny that there is a struggle for existence, and more importantly a struggle for reproduction. Malthus had left his mark. Few were going to deny that some will get through and some will not, that there will be differences between the successful and the unsuccessful, differences between the fit or fitter and the unfit or less fit. The big question was whether, even if natural selection were working flat out, it could have the effects that Darwin claimed it would, namely ongoing systematic change in the direction of ever-improved adaptations. And this of course brings us around to the variations on which natural selection supposedly works. Darwin admitted bluntly, "Our ignorance of the laws of variation is profound" (*Origin*, 167). Darwin did not know what causes variations, how frequent they are, what different kinds they are, and crucially what happens to variations during reproduction. Is the pattern of human sexuality the norm? Males and females produce males and females. Or is the pattern of human skin color the norm? One white and one black spouse produce intermediate kids, like President Obama. If the latter, and it really does seem that this is the common pattern, then however effective natural selection might be in one generation, then variations could well be blended out in two or three generations.

The Scottish engineer (and former classmate of Darwin) Fleeming Jenkin made the point with an example, uncomfortable for us, paradigmatically Victorian:

Suppose a white man to have been wrecked on an island inhabited by negroes. . . . Our shipwrecked hero would probably become king; he would kill a great many blacks in the struggle for existence; he would have a great many wives and children, while many of his subjects would live and die as bachelors. . . . Our white's qualities would certainly tend very much to preserve him to good old age, and yet he would not suffice in any number of generations to turn his subjects' descendants white. . . . In the first generation there will be some dozens of intelligent young mulattoes, much superior in average intelligence to the negroes. We might expect the throne for some generations to be occupied by a more or less yellow king; but can any one believe that the whole island will gradually acquire a white, or even a yellow population. . . ?

Jenkin concluded:

> Here is a case in which a variety was introduced, with far greater advantages than any sport every heard of, advantages tending to its preservation, and yet powerless to perpetuate the new variety.

Darwin, one might add, was not downcast by this argument. In the fifth edition of the *Origin*, he responded to this "able and valuable" article by agreeing that one or two variations would not do the job, but, if there were many variations, selection could have effects:

> If, for instance, a bird of some kind could procure its food more easily by having its beak curved, and if one were born with its beak strongly curved, and which consequently flourished, nevertheless there would be a very poor chance of this one individual perpetuating its kind to the exclusion of the common form; but there can hardly be a doubt, judging by what we see taking place under domestication, that this result would follow from the preservation during many generations of a large number of individuals with more or less curved beaks, and from the destruction of a still larger number with the straightest beaks. (*Origin*, 5th ed., 104–5)

It's all a little bit ad hoc, even though Darwin tried to give some theoretical backing to his assumption by introducing his "provisional hypothesis" of "pan-angenesis." Supposedly there are little gemmules (hypothetical tiny particles) all over the body; these get altered by external circumstances, then circulate through the body being collected in the sex cells, thus causing abundant variation in the next generation. No one was much impressed by this. Darwin himself, although he introduced it in his two-volume *Variation of Animals and Plants under Domestication*, kept it out of the later *Descent of Man*. In any case, it seems more a mechanism for supporting Lamarckism – the gemmules are altered as the blacksmith's arms get bigger and more muscular – than something speaking directly to the swamping problem of Fleeming Jenkin. There was an unsolved difficulty at the heart of Darwin's theory, and scientists knew this.

The swamping problem came from within. There was another problem, of a different ilk, that came from without. "I am greatly troubled at the short duration of the world according to Sir W. Thompson, for I require for my theoretical views a very long period *before* the Cambrian formation." This is an extract from a letter by Darwin to James Croll, January 31, 1869. He is referring to the fact that the physicist William Thomson, later Lord Kelvin, one of the leading scientists of the day, had used physical data – essentially treating the Earth as a cooling body that started as a mass of molten rock, estimating temperature gradients and so forth – to calculate the age of the Earth at a maximum of 20 to 400 million years, with 98 million as the best estimate. A long time – but hardly long enough for the slow process of Darwinian evolution through natural selection. In the first edition of the *Origin*, Darwin calculated that eroding the area between England's North and South Downs, chalky areas below London – "denudation of the Weald" – would have taken 300 million years. Later, under criticism, he withdrew that estimate, but it gives an idea of the difference between the kind of timespan that the evolutionist Darwin needed – the denudation would be just one small period in a total age of literally billions of years – and the kind of timespan that the physicists were prepared to allow him.

Over the years, Darwin and his fellow biologists and geologists wriggled under the limits supposedly set by physics. For instance, at the end of the 1870s, we find Charles Darwin rejoicing at the suggestion by his physicist son George that the tides caused heating through friction and so slowed the cooling of the Earth, thus throwing Kelvin's calculations off balance. Truly, though, matters rested until the beginning of the twentieth century when the discovery of radioactive decay and its warming effects showed that Kelvin's calculations were totally off the mark. Today, we think the Universe is 13.8 billion years old, the Earth 4.5 billion years old, and the start of life relatively shortly after that (3.8 billion years ago). There is quite enough time for natural selection to do its job. One should add that Kelvin lived to see his calculations overthrown, a result he accepted with fairly good grace, although he never publicly announced his retraction.

From our perspective about the existence or non-existence of the Darwinian Revolution, the damage was done. Then, as now, the simple fact is that physics had higher prestige than biology, and general opinion – within and without the scientific community – was that natural selection, demanding so much time, was simply not up to the job. Typical is the opinion of the American novelist Mark Twain.

Some of the great scientists, carefully ciphering the evidences furnished by geology, have arrived at the conviction that our world is prodigiously old, and they may be right but Lord Kelvin is not of their opinion. He takes the cautious, conservative view, in order to be on the safe side, and feels sure it is not so old as they think. As Lord Kelvin is the highest authority in science now living, I think we must yield to him and accept his views. (Mark Twain, *Letters from the Earth*.)

The Bowler thesis vindicated!

## Utility

A second point in favor of the Bowler thesis is that, given the kind of science that was then being done by professional scientists interested in evolutionary questions, natural selection really was not needed. If anything, it was a bit of a hindrance. Darwin admitted this in the *Origin*. A morphologist/paleontologist working out paths of evolution (phylogenies) searches for similarities of body structure, homologies. Secondary adaptations can only make this task more difficult. Birds and insects both have adaptations for flying, but you would not use this as evidence that birds and insects are more closely related than birds and mammals. In Owen's language, birds and mammals share the vertebrate archetype, and insects do not. Darwin wrote:

> On my view of characters being of real importance for classification, only in so far as they reveal descent, we can clearly understand why analogical or adaptive character, although of the utmost importance to the welfare of the being, are almost valueless to the systematist. For animals, belonging to two most distinct lines of descent, may readily become adapted to similar conditions, and thus assume a close external resemblance; but such resemblances will not reveal—will rather tend to conceal their blood-relationship to their proper lines of descent. (*Origin*, 427)

There speaks the man who had made his mark as a professional biologist through an eight-year study of barnacle morphology.

*Malgré lui*, Darwin set the pattern. Notably, Darwin's most fervent supporter, his "bulldog," had doubts. In an early review of the *Origin* – the day after

Christmas, December 1859, in the establishment newspaper *The Times* – Thomas Henry Huxley wrote:

> That this most ingenious hypothesis enables us to give a reason for many apparent anomalies in the distribution of living beings in time and space, and that it is not contradicted by the main phenomena of life and organization appear to us to be unquestionable; and so far it must be admitted to have an immense advantage over any of its predecessors. But it is quite another matter to affirm absolutely either the truth or falsehood of Mr. Darwin's views at the present stage of the inquiry.

He continued to doubt selection and, quite remarkably, in his very extensive series of university lectures on the life sciences, mentioned it virtually not at all. Huxley would have felt no need to present natural selection, either pro or anti. The simple fact is that the science that Huxley and so many of his fellow professionals were doing was not the kind of science that needed selection, at least not as a day-to-day tool. Like Darwin, they were first physiologists and embryologists and anatomists, working out the functioning and relationships of organisms, and secondarily paleontologists, working out the history of life.

This began a tradition lasting down to the present day. Fifty years ago, around 1970, there was a huge controversy in the area of biology dealing with classification – systematics. In the Anglophone world, at least, the standard approach was that of the Darwinian evolutionist. It is true that birds and crocodiles share common reptilian ancestors, and this should not be ignored when doing classifications. But let us not overdo it. We know now that humans and chimpanzees are more closely related than chimpanzees and gorillas, but – having regard to their adaptive natures – no classification should put humans with chimpanzees, and gorillas apart, rather than chimpanzees with gorillas, and humans apart. Then came challengers, fueled at least in part by the advent of computers, opening up the ready massage of huge amounts of data. *Pheneticism*, trying to classify on appearances alone – count the similarities and differences – was short-lived. Apart from anything else, what is an appearance? Does a bald man have but one distinguishing feature, or a thousand, for every missing hair? Far more successful and long-lived, indeed the dominant position today, was phylogenetic systematics or *cladism*, formulated and promoted by the German biologist Willi Hennig. Here, in the name of objectivity, phylogeny and phylogeny alone is what counts. If phylogeny shows that humans and chimpanzees should go together, then so be it. Gorillas are out in the cold. What counts in classification is when lines split apart. That gives you a handle with which to make your divisions into separate groups. Above all, you should not make emerging differences of appearance influence division-making. If, without speciation, the members of a group change from (say) egg laying to live births, they all remain members of the same group. In the words of American Museum of Natural History Curator, spider expert, Norman Platnick:

> To attempt to divide a species between speciation events would indeed be arbitrary: we would not call an individual person by one name at age 10 and a different name at age 30. Dividing species at their branching points, however, becomes not only non-arbitrary but necessary: we would not call a son by the same name as his father. (*Review*, 97)

The bottom line of interest to us is that causes are seen as irrelevant. Natural selection no longer has pride of place. According to philosopher David Hull:

> In general, cladists seem to be moving toward the position that the particulars of evolutionary development are not relevant to cladism. It does not matter whether speciation is sympatric or allopatric, saltative or gradual, Darwinian or Lamarckian, just so long as it occurs and is predominantly divergent... (*Limits of Cladism*, 418)

Do we go all the way and deny that classification is evolutionary? Do we say that it is simply a formal exercise to classify organisms, without reliance on or interest in how it all happened? There have been some – "pattern cladists" or "transformed cladists" – who have gone this far. Perhaps because he worked primarily on living organisms and not on fossils, ardent cladist Norman Platnick felt able to write: "If classifications (that is, our knowledge of patterns) are ever to provide an adequate test of theories of evolutionary processes their construction must be independent of any particular theory of process". In a similar vein, here is Colin Patterson: "We must remember the distinction between the cart – the explanation – and the horse – the data. And where models are introduced in phylogenetic reconstruction, we should prefer models dictated by features of the data to models derived from explanatory theories". Most would not go this far. Everything is a function of evolution, and it is disingenuous to pretend otherwise. As David Hull argued:

> Pattern cladistics has remained on the fringe because of, first, its implausible assumption that there can be pure observation untainted by theory; and second, its rejection of the evolution assumption. Few systematists now think that a classification not based on evolutionary branching and history has any real signification or justification. The developing consensus is that Darwin was right – a natural classification must be genealogical.

It is not very persuasive to say that one is not denying evolution, just discounting it, when one's thinking is evolutionary from the first.

## Organicism *Redivivus*

Let us move to a third support of the Bowler thesis. Given what we have seen, it is little surprise that in Germany, the home country of Immanuel Kant, a school of thought sprang up determined to resurrect the organic root metaphor as the correct way to conceptualize experience. The poet Johann Wolfgang von Goethe, the philosopher Friedrich Schelling, and the anatomist Lorenz Oken were leading figures in this "Romantic movement." They were open in their debts to the Greeks. Indeed, the teenage Schelling wrote a sixty-page essay on the *Timaeus*. Two things from Plato were of extreme importance. First, obviously, the organic model. Coming after two millennia of Christianity with its emphasis on origins, this pushed the Romantics towards evolutionism; not the harsh evolutionism of the Industrial Revolution – that of the two Darwins in particular – but more an internal growth. As the organism develops from forces within, so species evolve from forces within. The second debt was holism. Plato's theory of Forms, with the Form of the Good integrating all, was fundamental. This was epitomized by Richard Owen's theory of the vertebrate archetype, an idea he borrowed directly (without a great deal of acknowledgement) from the Romantics (see Fig. 1.7). This holism was in direct contrast to the reductionism of the mechanist, who breaks things down to see how they work – the springs and the cogs in the watch machinery, as in Charles

Darwin's view of organisms, broken down into their separate adaptations to see how they come together in the functioning of the whole.

The organicism of the Romantics was to have great influence, particularly on the mid-century English man of letters and ardent evolutionist Herbert Spencer. A slightly younger contemporary of Charles Darwin, Spencer became an evolutionist and started publishing on the topic early in the 1850s. In later years, privately he acknowledged explicitly his debt to Schelling, a debt to which he gave a characteristically idiosyncratic interpretation, incorporating English concepts and attitudes in a way alien to German Romanticism. Spencer, like Darwin, seized on Malthus, leading to some uncomfortably harsh views on society. In 1851, he wrote:

> Blind to the fact that under the natural order of things, society is constantly excreting its unhealthy, imbecile, slow, vacillating, faithless members, these unthinking, though well-meaning, men advocate an interference which not only stops the purifying process but even increases the vitiation—absolutely encourages the multiplication of the reckless and incompetent by offering them an unfailing provision, and discourages the multiplication of the competent and provident by heightening the prospective difficulty of maintaining a family. (*Social Statics*, 323–324)

By 1852, declaring himself to be an evolutionist – "Those who cavalierly reject the Theory of Evolution as not being adequately supported by facts, seem to forget that their own theory is supported by no facts at all" Spencer then went on to give a remarkable anticipation of natural selection.

> This truth we have recently seen exemplified in Ireland. And here, indeed, without further illustration, it will be seen that premature death, under all its forms, and from all its causes, cannot fail to work in the same direction. For as those prematurely carried off must, in the average of cases, be those in whom the power of self-preservation is the least, it unavoidably follows, that those left behind to continue the race are those in whom the power of self-preservation is the greatest—are the select of their generation. (*Theory of Population*, 500)

Selection, however, was never Spencer's real focus of change. He believed that the struggle leads to effort – like Margaret Thatcher, more than a century after him, he saw Malthusianism leading to the success of the talented rather than failure of the inadequate – and then, through Lamarckian inheritance, the individuals in a population will be changed and improved.

> So that, whether the dangers to existence be of the kind produced by excess of fertility, or of any other kind, it is clear, that by the ceaseless exercise of the faculties needed to contend with them, and by the death of all men who fail to contend with them successfully, there is ensured a constant progress towards a higher degree of skill, intelligence, and self-regulation—a better co-ordination of actions—a more complete life.

"Ceaseless exercise," "constant progress" – change coming from within in a totally organic manner. Fertility rates moderate, Spencer goes on to say, and population pressures decrease. With growth in brain size, fertility falls away and the Malthusian consequences are reduced. "Undue production of sperm-cells involves cerebral inactivity. The first result of a morbid excess in this direction is headache, which may be taken to indicate that the brain is out of repair; this is followed by stupidity; should the disorder continue, imbecility supervenes, ending occasionally in insanity" (493). Spencer himself, wisely avoiding headaches, had no children at all.

Later in the decade, explicitly reinforcing his holism, Spencer likened societies to organisms. And inspired by the second law of thermodynamics, Spencer believed that groups or societies would be in equilibrium, something disturbs them, and then they achieve equilibrium at a higher level. Dynamic equilibrium!

Spencer was not alone in his organic approach to evolution. In Germany, in the second half of the nineteenth-century, Ernst Haeckel promoted a vision of evolution that, for all he often pretended otherwise, owed more to the Romantics than to Darwin. Likewise in America. Vernon Kellogg (later to make waves with *Headquarters Nights*, a devastating critique of the German High Command during the Great War) was a long-time biology professor at Stanford University and was the author of *Darwinism Today*, published in 1905, a detailed study of the state of play in evolutionary circles at the beginning of the twentieth century. Darwin's theory of natural selection does not come out well. "The fair truth is that the Darwinian selection theories, considered with regard to their claimed capacity to be an independently sufficient mechanical explanation of descent, stand today seriously discredited in the biological world." What then is the alternative? Here, Kellogg is not quite as lucid as one might like. But it does seem that he endorses some kind of organicist process of directed – or quasi-directed – variations, with selection relegated to cleaning up after the real action is over. Why cannot the simple fluctuating or Darwinian variations be chiefly the result of the inevitable variations in the epigenetic (developmental) factors? It seems to be an exercise for the reader to ferret out how these epigenetic factors function. We shall see that there are those today who might have brought comfort to this puzzled biologist.

## Natural Selection Valued and Used

So much for the case against the flourishing of Darwin's primary cause of evolution, natural selection. Three strikes and you are out. We, however, are not playing baseball. We have seen but part of the story. Start with the place where you are going to find natural selection useful as a tool of understanding: the domain of fast-breeding organisms. Most obviously, for a nineteenth-century scientist, that means the world of insects. From the start, we find natural selection appreciated as a causal force behind design-like features, adaptations. Best known is the work on insect mimicry by Henry Walter Bates, a sometime traveling companion of Alfred Russel Wallace, as they collected butterflies along the Amazon in Brazil, who published this work in 1862–63. Bates, who entered into detailed correspondence with Darwin, argued that some insects (he focused on the butterfly *Leptalis*, now called *Dismorphia*) mimics other insects (in this case, the butterfly genus *Ithomia*). The latter are rejected by predating birds, because the plants on which they feast are foul-tasting. The *Leptalis*, to the contrary, are quite tasty, and heavily predated. They manage to survive and reproduce because they have lots of offspring, hoping that some at least will get through. There will, as always, be variation in the offspring, and Bates's claim was that those that do get through will on balance be closer in appearance to the foul-tasting *Ithomia* than the losers. Hence, mimicry.

> If a mimetic species varies, some of its varieties must be more and some less faithful imitations of the object mimicked. According, therefore, to

the closeness of its persecution by enemies, who seek the imitator, but avoid the imitated, will be its tendency to become an exact counterfeit, – the less perfect degrees of resemblance being, generation after generation, eliminated, and only the others left to propagate their kind. (*Contributions*, 512)

A year or two after, "Batesian mimicry" was paralleled by "Müllerian mimicry," where you have two distasteful species, and the members take advantage of being part of a large indistinguishable group. "Now if two distasteful species are sufficiently alike to be mistaken for one another, the experience acquired at the expense of one will likewise benefit the other; both species together will only have to contribute the same number of victims which each of them would have to furnish if they were different" (*Ituna and Thyridia*, 27).

Darwin was appreciative of this work, to the extent that he managed to get Bates a full-time job as secretary to the Royal Geographical Society. In a way, this was a bit of an (inadvertent) poison pill, because it meant that Bates was no longer doing creative science but administering an organization of upper-middle-class toffs. And this rather points to a general truth. The people who were interested in selection and fast-breeding organisms and similar work were not, at this time at least, full-time professional scientists, with university jobs. They were amateurs pursuing their hobby of butterfly collecting, often highly talented, but not working scientists like Huxley. Most remarkable of all was one Albert Brydges Farn, a civil servant and keen sportsman (that is, killer of birds) who, in 1878, wrote the most remarkable letter to Darwin about natural selection in observable action.

> My dear Sir,
>
> The belief that I am about to relate something which may be of interest to you, must be my excuse for troubling you with a letter.
>
> Perhaps among the whole of the British Lepidoptera, no species varies more, according to the locality in which it is found, than does that Geometer, *Gnophos obscurata*. They are almost black on the New Forest peat; grey on limestone; almost white on the chalk near Lewes; and brown on clay, and on the red soil of Herefordshire.
>
> Do these variations point to the "survival of the fittest"? I think so. It was, therefore, with some surprise that I took specimens as dark as any of those

in the New Forest on a chalk slope; and I have pondered for a solution. Can this be it?

It is a curious fact, in connexion with these dark specimens, that for the last quarter of a century the chalk slope, on which they occur, has been swept by volumes of black smoke from some lime-kilns situated at the bottom: the herbage, although growing luxuriantly, is blackened by it.

I am told, too, that the very light specimens are now much less common at Lewes than formerly, and that, for some few years, lime-kilns have been in use there.

These are the facts I desire to bring to your notice.

I am, Dear Sir, Yours very faithfully,

A. B. Farn

Letter from Albert Brydges Farn on November 18, 1878.

One might have thought that Darwin would at once have put Bates's work right into the *Origin* as an example of natural selection in action and have brought out a new edition of the *Origin* with Farn's letter reprinted across from the title page. Not at all. Bates made it into a later edition of the *Origin*, towards the end, and Farn seems not to have had a reply. The sneaking suspicion is that, as we have had reason to believe, as a scientist, Darwin was not much of a Darwinian! Apart from some experiments showing a struggle between plants to survive (*Origin*, 33), he never thought to go looking for natural selection in action or to work on it experimentally – or to subsidize others who might have done so. Darwin seems to have been convinced that selection is always so slow-working, one will never see it in significant action.

Later in the century, things did pick up a little. The entomologist J. W. Tutt in 1890 did pathbreaking work on industrial melanism – the sort of thing noted by Farn, where butterflies and moths acquire adaptive camouflage to protect them against changing backgrounds due to pollution by soot and the like (Fig. 4.3). "I believe . . . that Lancashire and Yorkshire melanism is the result of the combined action of the 'smoke,' etc., plus humidity [thus making the bark darker], and that the intensity of Yorkshire and Lancashire melanism produced by humidity and smoke, is intensified by 'natural selection' and 'heredity tendency'". Finally, we start to see the professionals coming on board.

Edward Bagnall Poulton, later to become Hope Professor of Zoology at the University of Oxford, was ever an ardent supporter of natural selection. In his *The Colours of Animals*, published in 1890, for instance, there is huge enthusiasm for Bates's explanation of insect mimicry. Darwin's secondary mechanism of sexual selection also came in for praise:

> Sexual Colours only developed in species which court by day or twilight, or have probably done so at no distant date. The appearance of beautiful colours and patterns, which are displayed in courtship, invariably occurs in diurnal or partially diurnal animals. The colours only appear when the conditions for female preference are present also. If we compare butterflies with moths, or moths which fly by day and twilight with those which fly in darkness, we find that brilliant tints and ornamental patterns are only found when there is light enough for the female to see them. (*Colours*, 331)

This, to be candid, is not quite what one expects from a "non-Darwinian Revolution." The same is true of the work of W. F. R. Weldon, towards the end of his rather short life as Linacre Professor at Oxford. He studied crabs at Plymouth, on the English south coast, finding that their frontal breadth was a function of the sediment in the water. Both by observation in nature as well as by experiments, he showed that the breadth is controlled by natural selection. Dirtier water means more clogging of the crabs' filtering apparatus, and hence a reduced frontal breadth is an adaptive advantage. Conversely, cleaner water means less clogging and the worth of an increased frontal breadth. The conclusion? To quote Weldon:

> I hope I have convinced you that the law of chance enables one to express easily and simply the frequency of variations among animals; and I hope I have convinced you that the action of Natural Selection upon such fortuitous variations can be experimentally measured, at least in the only case in which anyone has attempted to measure it. I hope I have convinced you that the process of evolution is sometimes so rapid that it can be observed in the space of a very few years. (*Monad to Man*, 231)

Finally, talking of science, selection studies may have had low academic status, but this is nothing to the problems of agriculture. It has difficulty edging above education in academic status and falls even below sociology. And yet, as you might have expected from a field that made such a contribution to the argument of the *Origin*, there was always interest in selection, natural as well as artificial. They are, in their way, all part of the same process. An instructive case is the entomologist and taxonomist John Henry Comstock, professor at Cornell University in Upstate New York. Conveniently, he founded his own publishing house and used it to give advice to young scientists. Classification, for instance, demands a steady approach, one step at a time. You begin with the single, isolated organ. As he wrote in 1893:

> First the variations in form of this organ should be observed, including paleontological evidence if possible; then its function or functions should be determined. With this knowledge endeavour to determine what was the primitive form of the organ and the various ways in which this primitive form has been modified, keeping in mind the relation of the changes in form of the organ to its functions. In other words, endeavour to read the action of natural selection upon the group of organisms as it is recorded in a single organ. The data thus obtained will aid in making a provisional classification of the group. (*Evolution and Taxonomy*, 41)

One would want to see more evidence of natural selection actually being used as a tool of understanding, but again there seems to be a more positive approach than Bowler leads us to expect.

## Natural Selection Out in Public

Move now from the world of science and ask about the fate of natural selection out in the world of regular human beings. The usual account starts (and often ends) with the story of the clash at the annual meeting in Oxford of the British Association between Thomas Henry Huxley and the Bishop of Oxford, William Wilberforce. Supposedly, Wilberforce asked Huxley if he was descended from monkeys on his grandfather's side or his grandmother's. Supposedly, Huxley responded that he had rather been descended from a monkey than from a bishop of the Church of England. Verbal fireworks, and no one much interested in Darwin's actual theory. Just a nasty aftertaste that it is clearly against religion.

This impression that there was little or no knowledge of, or interest in, not just evolution but Darwin selection-fueled evolution, could not be farther from the truth. Remember, in the years after the *Origin*, we are in a world that does not have either film or television. The printed word was all-powerful – magazines, novels, poetry. It was not at all uncommon – the Darwins were paradigms – for the whole family to gather round, perhaps in the evening, and have one member – it was Emma, Charles's wife, who did the honors for the Darwins – read aloud to the group. Charles Dickens's weekly magazine, *All the Year Round*, had a circulation of 100,000. Up to half a million were reading or hearing its contents. The magazine published two anonymous articles on the *Origin* in 1860 and another in 1861.

> How, asks Mr. Darwin, . . . have all these exquisite adaptations of one part of the organisation to another part, and to the conditions of life, and of one distinct organic being to another, been perfected? He answers, they are so perfected by what he terms Natural Selection – the better chance which a better organised creature has of surviving its fellows – so termed in order to mark its relation to Man's power of selection. Man, by selection in the breeds of his domestic animals and the seedlings of his horticultural productions, can certainly effect great results, and can adapt organic beings to his own uses, through the accumulation of slight but useful variations given to him by the hand of Nature. But Natural Selection is a power incessantly ready for action, and is as immeasurably superior to man's feeble efforts, as the works of Nature are to those of Art. Natural Selection, therefore, according to Mr. Darwin – not independent creations – is the method through which the Author of Nature has elaborated the providential fitness of His works to themselves and to all surrounding circumstances. (the author was David Thomas Ansted, a professional geologist)

Not much ambiguity here. Opponents are referred to as "timid," and the treatment overall is very positive:

> We are no longer to look at an organic being as a savage looks at a ship— as at something wholly beyond his comprehension; we are to regard every production of nature as one which has had a history; we are to contemplate every complex structure and instinct as the summing up of many contrivances, each useful to the possessor, nearly in the same way

as when we look at any great mechanical invention as the summing up of the labour, the experience, the reason, and even the blunders, of numerous workmen. (*Species*, 299)

Similar sentiments are expressed by E.S. Dixon in 1862, in a little tale told in a competing weekly, *The Cornhill Magazine* – also with a circulation of 100,000 – edited also by a novelist, William Thackeray. The theme is struggle and success and failure. Lest there be any doubt that this is about Darwin, the two main characters are named "Natural selection" with occupation "Originator of Species" (mother) and "Struggle for Existence" (son). Perhaps it is too early to say that Darwin is absolutely right? "Still the book has given me more comprehensive views than I had before ... Here we are offered a rational and a logical explanation of many things which hitherto have been explained very unsatisfactorily, if at all. It is conscientiously reasoned and has been patiently written. If it be not the truth, I cannot help respecting it as sincere effort after truth" (*A Vision of Animal Existences*, 318).

This is but the beginning. In the next fifty years, natural selection (and perhaps even more sexual selection) was used to frame plots and poems and much more. About George Eliot's greatest (and most-admired) novel, Henry James grumbled in 1873: "*Middlemarch* is too often an echo of Messrs. Darwin and Huxley" (*Middlemarch*, 428). But this is nothing compared with Eliot's later novel, *Daniel Deronda*, published in 1876. It is all about sexual selection and how choices can work, as they do for the selfless hero Daniel, or not work, as they do not for the selfish anti-heroine Gwendolen Harleth, who marries Mallinger Grandcourt for his money alone. "It is because I was always wicked that I am miserable now" (*Daniel Deronda*, 25).

Published a decade or so later, in 1891, George Gissing's novel *New Grub Street* is structured entirely around the workings of natural selection. The tale is of two prospective authors, one hugely talented but unable or unwilling to make compromises, and the other less talented but much more politically savvy.

You have no faith. But just understand the difference between a man like Reardon and a man like me. He is the old type of unpractical artist; I am the literary man of 1882. He won't make concessions, or rather, he can't make them; he can't supply the market. I— well, you may say that at present I do nothing; but that's a great mistake, I am learning my business. Literature nowadays is a trade. Putting aside men of genius, who may succeed by mere cosmic force, your successful man of letters is your skillful tradesman. He thinks first and foremost of the markets; when one kind of goods begins to go off slackly, he is ready with something new and appetising. He knows perfectly all the possible sources of income. (38–39)

As happens in these sorts of stories, the genius does not make it alive to the last chapter, while the practical chap ends with the editorship of a prized journal and marries the widow of his rival. Critics often complain that Gissing's heroes and heroines are less-than-worthy human beings, but that is to miss the point. In the Darwinian world, it is success that counts, not perfection. The person of vigor is the winner. As the twice-married wife knew well: "though she could not undertake the volumes of Herbert Spencer, she was intelligently acquainted with the tenor of their contents; and though she had never opened one of Darwin's books, her knowledge of his main theories and illustrations was respectable" (397).

Darwinism pops up in unexpected places. At the end of *Tarzan of the Apes*, published in 1912, Darwin enthusiast Edgar Rice Burroughs has Jane caught in the throes of sexual selection. Sensibly, she suppresses "the psychological appeal of the primeval man to the primeval woman in her nature" and, following her Darwinian nature, makes the wise decision to marry the apparent Lord Greystoke (William Cecil Clayton) instead of the unacknowledged true Lord Greystoke (Tarzan).

Did not her best judgment point to this young English nobleman, whose love she knew to be of the sort a civilized woman should crave, as the logical mate for such as herself?

Could she love Clayton? She could see no reason why she could not. Jane was not coldly calculating by nature, but training, environment and heredity had all combined to teach her to reason even in matters of the heart. (*Tarzan of the Apes*, 340)

Fortunately, in the next novel in the series, Jane sees reason to revise her decision, and we are all set for twenty-five sequels.

As icing on the cake, turn to Constance Naden and her lightweight frolic, written about 1885. Titled "Natural Selection," it is more about sexual selection.

> I HAD found out a gift for my fair,
> I had found where the cave men were laid:
> Skulls, femur and pelvis were there,
> And spears that of silex they made.
>
> But he ne'er could be true, she averred,
> Who would dig up an ancestor's grave—
> And I loved her the more when I heard
> Such foolish regard for the cave.
>
> My shelves they are furnished with stones,
> All sorted and labelled with care;
> And a splendid collection of bones,
> Each one of them ancient and rare;
>
> One would think she might like to retire
> To my study— she calls it a "hole"!
> Not a fossil I heard her admire
> But I begged it, or borrowed, or stole.
>
> But there comes an idealess lad,
> With a strut and a stare and a smirk;
> And I watch, scientific, though sad,
> The Law of Selection at work.
>
> Of Science he had not a trace,
> He seeks not the How and the Why,
> But he sings with an amateur's grace,
> And he dances much better than I.
>
> And we know the more dandified males
> By dance and by song win their wives—
> 'Tis a law that with avis prevails,
> And ever in Homo survives.

Shall I rage as they whirl in the valse?
Shall I sneer as they carol and coo?
Ah no! for since Chloe is false
I'm certain that Darwin is true.

(*Poetical Works*, 207–8)

Take note that women were contributing to this spread of Darwinian ideas. Take note also that women were excluded from professional science. Huxley would not allow women in his classes. Given the keen intellects at work, do not rush too quickly to label the non-professional contribution as "pop" or some other demeaning category. George Eliot and Constance Naden were up there with Henry Walter Bates. Let's have no more talk about "Non-Darwinian Revolutions."

# 4   The Synthesis

Now we come to the elephant in the room. Darwin's theory was incomplete. When the theory was completed, would natural selection prove to be that effective? Although he threw in a lot of assorted, presumed-relevant facts, no one, starting with Darwin, had much idea about the nature of variation – how it comes, what form it takes, how regular it is. And, without this knowledge, given that natural selection supposedly works on this variation, it is hard to make definite judgments about its effectiveness; especially since Darwin stressed that, although variation has causes, it is random in the sense of not appearing according to need. When he was not pushing the Lamarckian alternative, he was adamant that it is selection alone that is responsible for adaptation.

## Mendel

We have seen that critics were quick to point out that there were problems. One that many thought was definitive was based on the general belief that characteristics would, on balance, be blended from generation to generation. As you probably know, the solution to Darwin's problem was being dis-covered across Europe, in what was then the Austro-Hungarian Empire, by the friar Gregor Mendel. One should stress at once that Mendel had no real grasp of the implications of what he was doing – he had a copy of the *Origin* (in German), and his markings in its margins were mainly directed to the question of whether a Catholic priest could accept evolution (he decided he could). It was at the beginning of the twentieth century that three people, separately, realized the significance of Mendel's work, and then the story began to pick up.

Mendel, working on pea plants, provided the empirical basis for the hypothesis that physical characteristics like flower color are controlled by "factors" – units of inheritance, what we now call "genes." In organisms, these factors come in pairs, one inherited from each parent. Sometimes the factors are identical (known today as homozygotes) and sometimes different (heterozygotes). Which factor is transmitted from a parent to an offspring depends entirely on chance and is quite independent of the member of the pair transmitted from the other parent. Also, Mendel argued that where there are differences (heterozygotes), the effects of one factor could generally block out the effects of the other factor – that is, one was "dominant" and the other "recessive." All-importantly, as can be seen from the diagram (Fig. 4.1), there is no blending of features out of existence. The factors, the genes, go on unchanged, and this means that the physical characteristics (known today as the "phenotype") can keep reappearing unchanged. This doesn't mean that you never get blending. Sometimes, neither gene is dominant and the offspring have blended physical effects; but then, in the next generation, with similar matching up to similar, the original phenotypes reappear. The worries of someone like Fleeming, who thought "once blended, always blended," are assuaged. (These potentially interacting genes are known as "alleles.")

The English biologist William Bateson, who had already interested himself in the nature of variations, picked up on this new approach and became its champion. However, going against the professional supporters of natural selection (particularly Weldon, who incidentally had been his teacher), Bateson presented Mendelism as an alternative – if not actually refuting the existence of natural selection, at the least rendering it superfluous. He thought the factors alone were adequate to support change. In his clarion call, *Mendel's Principles of Heredity: A Defence*, published in 1902, the chip on Bateson's shoulder is there from the start:

> If species had really arisen by the natural selection for impalpable differences, intermediate forms should abound, and the limits between species should be on the whole indefinite. As this conclusion follows necessarily from the premises, the selectionists believe and declare that it represents the facts of nature. Differences between species being by axiom indefinite, the differences between varieties must be supposed to be still less definite. Consequently the conclusion that evolution must proceed by

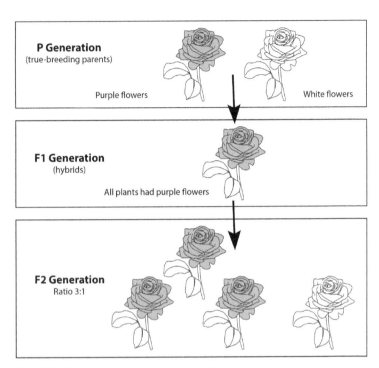

**Figure 4.1**    One parent has two purple-causing alleles and the other two white-causing alleles. If purple is dominant over white, the offspring are purple. But then, among their offspring, white organisms will reappear. (The ratio in generation 2, known as F2, will be 3:1 purple to white, because, on average, you will get one plant with two purple-causing genes, hence physically purple; two plants with one purple- and one white-causing gene, hence physically purple; and one plant with two white-causing genes, hence physically white.)

insensible transformation of masses of individuals has become an established dogma. Systematists, entomologists or botanists for example, are daily witnesses to variation occurring as an individual and discontinuous phenomenon, but they stand aside from the debate; and whoever in a discussion of evolutionary theory appeals to the definiteness of varietal distinctions in colour for instance, or in form, as recognizable by common observation without mechanical aid, must be prepared to meet a charge of want of intelligence or candour. (3)

Later, spelling things out:

There is also nothing in Mendelian discovery which runs counter to the cardinal doctrine that species have arisen "by means of Natural Selection, or the preservation of favoured races in the struggle for life," to use the definition of that doctrine inscribed on the title of the Origin. By the arbitrament of Natural Selection all must succeed or fail. Nevertheless the result of modern inquiry has unquestionably been to deprive that principle of those supernatural attributes with which it has sometimes been invested. The scope of Natural Selection is closely limited by the laws of variation. (289)

Natural selection seems to be cleaning up after the party is over.

## Towards Population Genetics

The lines were drawn, even though – owing to Weldon's sudden death in 1906 – the initial rancor faded somewhat. Things did not stand still. Keeping our eye on the fate of natural selection, several factors moved scientific thinking forward. Most important, unraveling the nature of the units of heredity, the "genes," was the work at Columbia University in the second decade of the century by Thomas Hunt Morgan and his associates. They showed that the genes are physical things, to be found along the chromosomes, string-like entities in the centers (nuclei) of complex cells. They showed also that the chromosomes are paired, so genes come in pairs, called (as we know) "alleles" (or allelomorphs), with the possibility of identical pairs (homozygotes) or different pairs (heterozygotes). Genes can be altered, spontaneously as it were – "mutations" – and, crucially important, genes can have very small effects which can be combined in an additive manner. As the eminent statistician Ronald Fisher put it in 1930: "The apparent blending in colour in crosses between white races of man and negroes is compatible with the view that these races differ in several Mendelian factors, affecting the pigmentation" (Genetical Theory, 17–18).

Another very significant move for our story comes from the fact that natural selection is a population-based causal process, as opposed to Lamarckism (which, into the new century, was coming under increasing attack), which is an individual-based causal process. Crucial therefore was the extension of genetics to populations. The all-important breakthrough came in 1908, thanks to the independent work of two thinkers, in England the mathematician G. H. Hardy and in Germany the physician Wilhelm Weinberg. What came to be known as the Hardy–Weinberg law showed that, without external disturbing factors, the genes (alleles) within a population will immediately move to a state of equilibrium: that is, the genetic variation in the population will remain constant (Fig. 4.2). Contrary to the assumption (one clearly behind the critique of Fleeming Jenkin) that minority forms will gradually be eliminated in a population, in large populations such forms will continue to exist, perpetually. This is an equilibrium law against which disturbing factors can now be introduced and their effects measured and studied.

Things came together around the end of the third decade of the century, between 1930 and 1932. It was then that the "population geneticists" presented their thinking to the world, in significant respects synthesizing the Darwinian and the Mendelian contributions into one united whole. Three men are usually picked out – in Britain, Ronald A. Fisher and J. B. S. Haldane,

---

Assume that we have a (large) population, with just two alleles in ratio $p$ to $q$. Hence, by definition:

$$p + q = 1$$

Given random mating, the Hardy–Weinberg law states that, in subsequent generations, the ratio of alleles will remain constant, and the ratio of genotypes will be:

$$p^2 + 2pq + q^2 = 1$$

$p^2$ is the frequency of individuals with the homozygous dominant genotype
$2pq$ is the frequency of individuals with the heterozygous genotype
$q^2$ is the frequency of individuals with the homozygous recessive genotype
("Genotype" is the genetic equivalent of the physical "phenotype".)

Figure 4.2 Hardy–Weinberg equilibrium

and in America, Sewall Wright. I will focus on the more influential two, Fisher and Wright. Following their work, we have the experimentalists who started to put empirical flesh on the theoretical frameworks, specifically E. B. Ford and his school of "ecological genetics" in Britain, and the Russian-born Theodosius Dobzhansky and associated following in the United States. The British label for the Darwin–Mendel unification is usually "neo-Darwinism"; the American label is "the synthetic theory" of evolution. We shall come down roughly to 1959, the hundredth anniversary of the *Origin*, when evolutionary studies could justifiably be said to have found its "paradigm."

## R. A. Fisher

The British way of conceptualizing things was to think in terms of large populations of organisms, and hence large "gene pools." As inheritance was particulate, without outside interference you were going to get equilibrium. As Fisher wrote in 1930: "The particulate theory of inheritance resembles the kinetic theory of gases with its perfectly elastic collisions, whereas the blending theory resembles a theory of gases with inelastic collisions, and in which some outside agency is required to be continually at work to keep the particles astir" (*Genetical Theory*, 11). This means, as Fisher stressed, that although mutations are going to be the building blocks of evolution, as it were, we must grasp from the beginning that it will not be a mutation that directs the course of evolution. It is natural selection or nothing. As Fisher also wrote in 1930:

> The whole group of theories which ascribe to hypothetical physiological mechanisms, controlling the occurrence of mutations, a power of directing the course of evolution, must be set aside, once the blending theory of inheritance is abandoned. The sole surviving theory is that of Natural Selection, and it would appear impossible to avoid the conclusion that if any evolutionary phenomenon appears to be inexplicable on this theory, it must be accepted at present merely as one of the facts which in the present state of knowledge seems inexplicable. The investigator who faces this fact, as an unavoidable inference from what is now known of the nature of inheritance, will direct his inquiries confidently towards

> a study of the selective agencies at work throughout the life history of the group in their native habitats, rather than to speculations on the possible causes which influence their mutations. (21)

Although Fisher does not make explicit reference to the Hardy–Weinberg equilibrium, his discussion makes it clear that it is this kind of thinking that is behind his overall vision. One thing that Fisher stressed is that mutations with small effects could in the long run be as effective as mutations with larger effects.

> If a change of 1 mm. has selection value, a change of 0.1 mm. will usually have a selection value approximately one-tenth as great, and the change cannot be ignored because we deem it inappreciable. The rate at which a mutation increases in numbers at the expense of its allelomorph will indeed depend on the selective advantage it confers, but the rate at which a species responds to selection in favour of any increase or decrease of parts depends on the total heritable variance available, and not on whether this is supplied by large or small mutations. (16)

So taken was Fisher with the analogy from physics (kinetic theory of gases), where the second law of thermodynamics states that entropy always increases (the disorder increases, meaning that the available usable energy is always on the decrease), that he came up with an equivalent for his version of population genetics. If you have variation in a population, then selection is always going to be pushing the group to the most efficient adaptation. Understanding by "fitness" the ability to reproduce compared to that of competitors, and by "variance" the difference from the norm, then Fisher's "Fundamental Theorem of Natural Selection" states: "The rate of increase in fitness of any organism at any time is equal to its genetic variance in fitness at that time." It does not of course follow that a population is always on the up. If "an organism be really in any high degree adapted to the place it fills in its environment, this adaptation will be constantly menaced by any undirected agencies liable to cause changes to either party in the adaptation."

Fisher continued:

> As to the physical environment, geological and climatological changes must always be slowly in progress, and these, though possibly beneficial

to some few organisms, must as they continue become harmful to the greater number, for the same reasons as mutations in the organism itself will generally be harmful. For the majority of organisms, therefore, the physical environment may be regarded as constantly deteriorating, whether the climate, for example, is becoming warmer or cooler, moister or drier, ... Probably more important than the changes in climate will be the evolutionary changes in progress in associated organisms. As each organism increases in fitness, so will its enemies and competitors increase. (45)

## Ecological Genetics

Fisher was the theoretician. It was the Oxford-based naturalist E. B. "Henry" Ford who – with his young associates – put empirical flesh on the theoretical skeleton. As might be expected, given the renewed emphasis on natural selection, there is significant continuity with the nineteenth century. One topic that received considerable attention was that of melanism. Philip M. Sheppard, a student of Ford, discussed the subject carefully in his *Natural Selection and Heredity*. He reported on experiments by H. B. D. Kettlewell, who released different-colored moths in appropriately different situations. "He found, both by the proportion of the two types recaptured and by direct observation of the birds taking the moths from the trunks, that carbonaria was far less heavily predated than the typical form. In one experiment, of equal numbers of the two forms 43 typical were taken to only 15 carbonaria. Consequently, the melanic was at a great advantage, which explains why it has become so common in polluted areas" (74). In the fifty years since this was written, experiment after experiment has confirmed this finding. Thanks to clear air laws, the melanic forms are now becoming less and less common. Trees are far less heavily polluted. (See Fig. 4.3)

Some of the most interesting selection work has focused on polymorphism, where one has two or more variants in a species. This, in itself, is no reason for special interest. If, for instance, a new mutant form proves superior to already existing forms, one expects polymorphism as the species changes from one type to the other. What is interesting is where it seems that selection maintains

**Figure 4.3** Industrial melanism. The (newer) darker form is camouflaged against the dirty tree whereas the (older) lighter form is not.

polymorphism. One case is known as "balanced heterozygote superior fitness." Ford seized on this straightaway. "Polymorphism will result if the heterozygote possesses some physiological advantage as opposed to the homozygote; it may for example be more fertile." The most famous recorded instance occurs in humans. In parts of Africa, a proportion of children are born with a fatal form of anemia. It is now known that there are two alleles involved: the "normal" allele and the "sickle-cell" allele, which in homozygotes causes the red blood cells to collapse into a sickle-type shape, functioning very inefficiently. These latter, about 4% of the population, die in early childhood. The reason the sickle-cell allele persists in populations is that such populations live in areas heavily infested by mosquitoes leading to malaria. Homozygotes for the normal allele are at great risk from malaria. However, heterozygotes for normal and sickle-cell alleles have an immunity to the malaria parasite, to the extent that they are twenty-six times fitter than normal (malaria-unprotected) homozygotes. The sickle-cell anemic children, who will not reproduce, are the cost of keeping the heterozygotes much better protected from the malaria parasite.

Another reason for polymorphism is selection for rareness. If, for instance, a predator must learn to recognize its prey, then the rare form in a group will be less likely to be predated. But as selection works in its favor, it will increase in number until the predators can recognize it as easily as the other form, and so the two forms will persist in a balance. The snail *Cepaea* has different forms of shell patterns. Thrushes eat these snails, and if one form is at an adaptive advantage – it blends more easily into its background so the thrushes do not see it – this form will increase in number against the other forms. Eventually, it will become so common (and the other forms relatively uncommon) that if the thrushes do not learn to recognize them, they will starve. Hence, now there is going to be selective pressure on the hitherto-advantaged form, and so at some point there will be equilibrium, with none of the forms having an edge over others.

The finding that natural selection can maintain variation in populations is of far greater significance than might appear at first sight. One of the main worries about natural selection is that, if the variations are random in the sense of non-directed, then it seems implausible that if an organism has reason to evolve (a new predator for instance, or an environmental change like an ice age), an appropriate variation will come along ready to be used. That is rather like having to write an essay on dictators and relying on the offerings of the Book of the Month Club. By the time anything vaguely useful came along, the course would be over and you would have failed. But now it seems that you have a library at your disposal. If it doesn't have a book on Hitler, then perhaps one on Stalin, and if not Stalin then how about Mao? There is almost certainly going to be something. It is true that there is going to be no one favorable option, but that is as it may be. A new predator? Then perhaps there are a range of genes for coloration, and one proves particularly useful for camouflage. Instead of the usual off-white, the move will be to a preponderance of dark grey organisms, who show up much less clearly against the vegetation. Or perhaps there will be genes for vile taste, so the predators soon learn to avoid you. Then again, there might be genes that make the organism favor a nocturnal existence. And so the story goes, with perhaps genes that say "Get the hell out of here fast." The whole point about natural selection is that it is all relative. There is no one unique solution. What works is what works.

## Sewall Wright

Time now to cross the Atlantic and look at the work and influence of Sewall Wright. He went as a graduate student to Harvard, and then was employed by the United States Department of Agriculture, where he worked on the breeding and improvement of shorthorn cattle. Wright was always interested in evolution, and this picked up when he moved to a lifelong position at the University of Chicago. Fisher and Wright communicated about technical mathematical issues, and they agreed about the mathematics as such. However, Wright's thinking led him in a very different direction from that of Fisher. Not at first. Like Fisher, he began with the Hardy–Weinberg equilibrium, using this as the background foundation. He wrote in 1931: "The starting point for any discussion of the statistical situation in Mendelian populations is the rather obvious consideration that in an indefinitely large population the relative frequencies of allelomorphic genes remain constant if unaffected by disturbing factors such as mutation, migration, or selection."

Wright then introduced the factors that might take a population from equilibrium: mutation, migration, and selection. He stressed the holistic nature of the last. "Selection, whether in mortality, mating or fecundity, applies to the organism as a whole and thus to the effects of the entire gene system rather than to single genes. A gene which is more favorable than its allelomorph in one combination may be less favorable in another" (101). He agreed with

Fisher in that selection can promote, as well as eliminate, diversity. With Fisher and Ford, Wright made mention of balanced heterozygote fitness, where the heterozygote is fitter than either homozygote, hence keeping different alleles in the same population. "There may be equilibrium between allelomorphs as a result wholly of selection, namely, selection against both homozygotes in favor of the heterozygous type" (102).

Then comes the decisive break, or innovation: "genetic drift" as a function of small populations.

> There remains one factor of the greatest importance in understanding the evolution of a Mendelian system. This is the size of the population. The constancy of gene frequencies in the absence of selection, mutation or migration cannot for example be expected to be absolute in populations of limited size. Merely by chance one or the other of the allelomorphs may be expected to increase its frequency in a given generation and in time the proportions may drift a long way from the original values. (106)

Why is this so very important? The answer comes most clearly in a follow-up paper that Wright wrote for a congress in 1932. First, through what has come to be known as (the metaphor of) an "adaptive landscape," Wright showed how he considered the course of evolution must follow (Fig. 4.4). For Fisher, thinking in terms of selection working within large populations, the whole group moves in the direction of what is of the highest adaptive value. As we have seen, Fisher did not think matters would ever be that stable; they would be open always to ongoing change, because of environmental changes and so forth. Wright seems to have thought more in terms of stable adaptive "peaks" on which groups find themselves, and the problem now becomes one of moving to a higher peak. For Fisher, selection does this on its own, for, in a way, peaks – points of highest adaptive value – are being created as the population changes in response to selection. For Wright, between the independently existing peaks there are going to be non-adaptive valleys. The problem is how to go down into these valleys so that one can then climb up to higher peaks. This is just what selection cannot do.

But genetic drift can! If a large population is divided into many subpopulations, with gene ratios drifting – that is, in a non-selection-driven manner – you are going to end up with many subpopulations with different non-adaptive

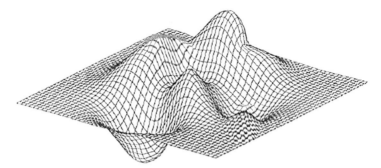

**Figure 4.4**   Adaptive landscape. The peaks of mountains represent areas of high fitness, the valleys areas of low fitness. Suppose a group of organisms were at the top of the lower peak – say, their color is fairly good at camouflage but not great. How could they ever improve their fitness to reach the top of the higher peak – changing their color to something different which is better camouflage – if it means going through changes of color along the way that offer very poor camouflage – going down into a valley in order to get to the foothills of the higher peak? Genetic drift can do the job, because it is not constrained by fitness demands. It is not natural selection.

genes fixed in these various subpopulations. They will, as it were, have slipped their populations down from the peak into a valley. Then, rather like individual mutations, one or more will happen to have just what is needed to climb back up to a different higher peak. Either this population spreads its innovative, now-adaptive (for a different peak) genes to other populations that can then follow, or it wipes out the competing, now-inferior other subpopulations. As Wright wrote in 1932: "The direction is largely random over short periods but adaptive in the long run" (*Roles of mutation*, 150). To elaborate: "let us consider the case of a large species which is subdivided into many small local races, each breeding largely within itself but occasionally crossbreeding. The field of gene combinations occupied by each of these local races shifts continually in a nonadaptive fashion (except in so far as there are local differences in the conditions of selection)." Wright continues:

> With many local races each spreading over a considerable field and moving relatively rapidly in the more general field about the controlling peak, the chances are good that one at least will come under the influence

of another peak. If a higher peak, this race will expand in numbers and by crossbreeding with the others will pull the whole species toward the new position. The average adaptiveness of the species thus advances under intergroup selection, an enormously more effective process than intragroup selection. (150)

Evolution has occurred – although, note, hardly Darwinian evolution. The major changes are non-adaptive, coming through drift, and only when the creative work has been done, with one or more populations now ready to climb a new peak, does some form of intergroup selection kick into action. One can start to sense why it was that, while the English called the Darwin–Mendel synthesis by the name "neo-Darwinism," the Americans did not want to be thus labelled, and the revised theory became known as the "synthetic theory." Overall, we have equilibrium, followed by disruption of equilibrium, followed in turn by achieving a new, higher equilibrium. And if this does not ring a bell, then, whatever your age, you are having a "senior moment." This is pure Herbert Spencer (see Chapter 3). The crucial creative steps are non-Darwinian, and the overall picture is one of dynamic or moving equilibrium.

Can this possibly be so? One of the founders of modern evolutionary thinking was no Darwinian? Someone who discounted the significance of natural selection? It is indeed true. When Sewall Wright was at Harvard, the inspiration of the biology department was Herbert Spencer –no great surprise, given the huge influence Spencer had had on late-nineteenth-century American thinking. Particularly important was the biochemist L. J. Henderson, who in 1917 wrote of Spencer that he was more a visionary than an empirical scientist, but "his generalizations, regarded as provisional and tentative hypotheses, possess genuine importance" (*Order of Nature*, 124). Wright was taught by Henderson and fell under his spell. He wrote to his brother Quincy, "I was always very much impressed with Henderson's ideas," and, acknowledging explicitly the direct influence back to Spencer, "I found him a very stimulating lecturer and got lots of ideas from him, 'condition of dynamic equilibrium' etc" (quoting unpublished letters). Not just this; as later writings show, the Spencer/Henderson holism also. In Wright's vision of evolution, the "shifting balance theory," when selection does then kick in, it is a group selection not an individual selection. As he wrote in 1945: "selection between the genetic systems of local populations of a species ... has been perhaps the greatest creative factor of all in making possible selection of genetic systems as wholes in place of mere selection according to the net effects of alleles" (*Tempo*, 416).

To make the picture complete, one might add that supplementing all of this was the influence of the French vitalist Henri Bergson (author of *L'évolution créatrice*), who had debts to Herbert Spencer and (with Spencer) to Friedrich Schelling. Wright himself acknowledged:

> The present discussion has dealt with the problem of evolution as one depending wholly on mechanism and chance. In recent years, there has been some tendency to revert to more or less mystical conceptions revolving about such phrases as "emergent evolution" and "creative evolution." The writer must confess to a certain sympathy with such viewpoints philosophically but feels that they can have no place in an attempt at scientific analysis of the problem. (154)

This did not stop Wright from taking the sting out of what he had just written by adding that a mechanistic explanation is at best just a surface explanation. "One may recognize that the only reality directly experienced is that of mind, including choice, that mechanism is merely a term for regular behavior, and that there can be no ultimate explanation in terms of mechanism—merely an analytic description."

"The only reality directly experienced is that of mind." Schelling – who famously wrote, "Nature is visible spirit, spirit is invisible nature" – would have hugged him.

## Fruit Flies

Theodosius Dobzhansky's *Genetics and the Origin of Species* (first edition published in 1937) was arguably the most important – certainly the most influential – book on evolutionary theory in the twentieth century. (This is even more true if you add in a dazzling series of papers on and around the topics of the book.) Dobzhansky rather set the scene by quoting in a disinterested fashion from two different perspectives, one (Fisher's) that makes selection all-important, and the other from two contemporary writers that do not: "We do not believe that natural selection can be disregarded as a possible factor in evolution. Nevertheless, there is so little positive evidence in its favor ... that we have no right to assign to it the main causative role in evolution" (151). In this mode, Dobzhansky runs through some of the traditional evidence in favor of selection and in support of its importance. Melanism gets short shrift. "Unfortunately the work on the industrial melanism has been restricted mainly to collecting the records of the happenings as they occur, and the causal analysis has lagged far behind, except for more or less gratuitous speculations" (160). Mimicry comes off little better. "Taken as a whole, an unprejudiced observer must, I think, conclude that an experimental foundation for the theory of protective resemblance is practically non-existent" (164). Dobzhansky was not entirely negative. "The adaptive value of the development of a longer pelage [furry coat] and a greater amount of wool in a cool climate is indeed obvious" (171). Dobzhansky was no Lamarckian, so it is selection at play here: "in a cold climate natural selection must favor the genotypes which, other conditions being equal, produce a warmer pelage." Overall, though, this is hardly the basis for a fully adequate theory of evolution.

Wright to the rescue! Dobzhansky, through a study of snails on oceanic islands, convinced himself that the variation between groups is non-adaptive: "The difficulty of proving that a given trait has not and never could have had an adaptive significance is admittedly great; nevertheless, the facts at hand are explicable, without stretching any logical point on the assumption that racial differentiation is due to mutations and random variations of the gene frequencies in isolated populations" (136). This set the scene for action by Wright's hypothesized scenario: "a colony that has reached a gradient leading to a new peak may climb it rapidly, increase in size, and either supplant the old species, or, more likely, form a new one that owes its allegiance to a new peak." Selection has a secondary, mopping-up role. "Natural selection will deal here not only with individuals of the same population (intra-group selection) but also, and perhaps to a great extent, with colonies as units (inter-group selection)" (190). In passing, remember that this latter process, inter-group selection, is something that Darwin explicitly eschewed. Darwin, with his division-of-labor thinking, would be loath to think that there could be no adaptive differences between separate populations.

This latter issue – adaptive differences between separate populations – was something that led to a significant change in Dobzhansky's thinking. The third edition of *Genetics and the Origin of Species* (1951) is very different from the first edition, being far more selection-friendly or Darwin-friendly. Dobzhansky's work on fruit flies, *Drosophila*, often in collaboration with Wright (who did the mathematics), had shown unambiguously that selection plays an active role in the life-cycles of small isolated populations. A 1943 paper by Dobzhansky (alone) on *Drosophila obscura* showed very significant genetic fluctuations that were clearly a function of ecological conditions at different times of the year. What might be behind all of this? A "plausible view is that Standard is favored while the populations are at their highest density levels in the year, and that Chiricahua [a form of fruit fly contrasted with Standard] is favored while the populations are dwindling toward their summer eclipse stage, presumably because of a relative food scarcity" (322).

Moving to the third edition of *Genetics and the Origin of Species*, when (to be fair) Ford and his group were in high gear, mimicry got a far friendlier treatment.

> There can no longer be a reasonable doubt that many animals are camouflaged in their natural surroundings. It is of course, a different problem whether the camouflage has developed under the influence of natural selection, because of the protection from enemies which these properties confer on their carriers. Here one can proceed only by inference, with experiments pointing the way. Such experimental data as are available support the natural selection theory. (102)

Melanism likewise:

> According to Ford's ingenious hypothesis, the spread of melanic mutants was precluded before the advent of industrial developments owing to the destruction of such mutants by predators, since dark individuals are not protectively colored. This disability is removed in the industrial regions by the general darkening of the landscape. The superior viability of the melanics is able, then, to assert itself, and their rapid increase in frequency is the result. (132–3)

While Wright's theorizing about evolution is not dropped, the endorsement is nothing like as positive. Selection is given a much larger role. "It is not possible at present to reach definitive conclusions regarding the role played in genetic drift in evolutionary processes" (164). That about says it all. There is far from total unanimity, but as we approach the hundredth anniversary of the *Origin*, in 1959, natural selection is in better and better shape. This being so, it is time to look philosophically at the concept itself.

# 5   Is Natural Selection a *Vera Causa*?

Time to pull back and get a little more conceptual. We need to ask some penetrating questions about the nature, the scope, the truth-value of natural selection. Finding answers, the quest begins in the past. Charles Darwin was a graduate of the University of Cambridge. The greatest British scientist of them all, Isaac Newton, was also a graduate of the University of Cambridge, and his spirit, his achievements, his reputation, infused every discussion about science, including about the life sciences. In his *Principia*, Newton started with his three laws of motion, together with his law of gravitational attraction, and then went on to infer, deductively, the pertinent terrestrial laws, those of Galileo, and the pertinent celestial laws, those of Copernicus affirming the heliocentric nature of the universe and those of later thinkers, especially Kepler on planetary motion. It was a given that the ambitious young Charles Darwin would want to show Kant dead wrong. There could be a Newton of the blade of grass, and that Newton was going to be Charles Darwin.

## Natural Selection as a *"Vera Causa"*

What truly was Newton's greatest achievement? It was to posit a force, a cause – gravitational attraction – that would explain how everything worked. It is for this reason that, having become an evolutionist in the early months of 1837, some six months after the end of the *Beagle* trip, Darwin then set off on an eighteen-month question to find the cause, the biological equivalent of gravitational attraction. This, of course, was natural selection. Did Darwin think he had been successful? He certainly did! On May 1863, he wrote to a George Bentham, the nephew of the philosopher Jeremy Bentham:

> In fact the belief in natural selection must at present be grounded entirely on general considerations. (1) on its being a vera causa, from the struggle for existence; & the certain geological fact that species do somehow change (2) from the analogy of change under domestication by man's selection. (3) & chiefly from this view connecting under an intelligible point of view a host of facts. (Letter from Darwin to George Bentham May 22, 1863)

A *vera causa* – a true cause! The canonical definition/characterization of this notion is to be found in the 1785 writings of the Scottish philosopher Thomas Reid.

> The first rule of philosophising laid down by the great Newton is this : – "*Causas rerum naturalium non plures admitti debere quam quae et vera sint et earum phenomenis explicandis sufficient*" – "No more causes, nor any other causes of natural effects, ought to be admitted, but such as are both true, and are sufficient for explaining their appearances." This is a golden rule; it is the true and proper test, by which what is sound and solid in philosophy may be distinguished from what is hollow and vain. (*Essays*, 34)

More immediately, Darwin was influenced by the two leading philosophers of science of his day, both Cambridge men: the astronomer John F. W. Herschel, and the general man of science and future Master of Trinity College, William Whewell. On Whewell's recommendation, after graduating and before going on the *Beagle* voyage, he read Herschel's *Preliminary Discourse on the Study of Natural Philosophy*, and, apart from ongoing mentorship by Whewell before and after the voyage, in the year of its publication (1837), twice (once quickly and once more slowly), he read Whewell's *History of the Inductive Sciences*. Although Darwin did not read Whewell's *Philosophy of the Inductive Sciences* (published in 1840), he read a lengthy and detailed review by Herschel. Both Whewell and Herschel were committed to what today we call a "hypothetico-deductive" view of scientific theories – axiomatic, with laws of nature as premises – as we find in the *Principia*. The force of gravitational attraction is a *vera causa* because it is the cause found in the axioms from which all else follows – "true and sufficient for explaining appearances." This is precisely the claim by Darwin about natural selection. It is a *vera causa* "from the struggle for existence; & the certain geological fact that species do somehow change."

Three points. First, as already noted, nothing in Darwin's work is particularly formal. The inference from the struggle to selection gets reasonably close; but, after that, "explanation sketch" is a generous way of putting things. Darwin sees this as working out the details and not as something requiring a radical new approach. Second, gravity occurs right up in one of the introductory axioms, whereas Darwin starts with the struggle and then goes on to selection. Darwin would not have thought this significant, nor indeed should he have done so. Third, for all their friendship and respect for each other's work, philosophically Herschel and Whewell were poles apart. Herschel, in the tradition of Hume, was an empiricist – start with the real world and go from there to hypotheses and the like. Whewell, in the tradition of Descartes, was a rationalist – start with the hypotheses and explain the world from there.

## Empiricist Analogy

Let us begin with Herschel. Along with the *Personal Narrative* of Alexander von Humboldt, whose natural history was a great influence on Darwin when the *Beagle* visited Brazil, Darwin said of Herschel's *Preliminary Discourse*: "No one or a dozen other books influenced me nearly so much as these two." What did he find in Herschel? Analogy! If we are after causal understanding, then we must start with the known, the experienced, and work to the unknown, the inferred. As Herschel wrote in 1830:

> If the analogy of two phenomena be very close and striking, while, at the same time, the cause of one is very obvious, it becomes scarcely possible to refuse to admit the action of an analogous cause in the other, though not so obvious in itself. For instance, when we see a stone whirled round in a sling, describing a circular orbit round the hand, keeping the string stretched, and flying away the moment it breaks, we never hesitate to regard it as retained in its orbit by the tension of the string, that is, by a force directed to the centre; for we feel that we do really exert such

a force. We have here the direct perception of the cause. When, therefore, we see a great body like the moon circulating round the earth and
not flying off, we cannot help believing it to be prevented from so doing,
not indeed by a material tie, but by that which operates in the other case
through the intermedium of the string,—a force directed constantly to the
centre. (*Preliminary Discourse*, 132)

By way of contemporary example, Herschel referred to Charles Lyell's magis-
terial three-volume *Principles of Geology* (1830–33) – the first volume of
which, remember, Darwin took with him on the *Beagle*, the others being
sent out to him in South America. Lyell was (what Whewell called)
a "uniformitarian," in seeing geology being a matter of regular laws and causes
working non-stop indefinitely, in a steady state, rather than a "catastrophist,"
seeing a directed process sometimes radically disturbed by major events from
outside. Lyell was worried by the fact that the fossil record shows that the
climate was different in the past, formerly (as in Paris and shown by palm trees)
much warmer, pointing to (as argued by the catastrophists) a decline in the
heat of the planet. Lyell wanted everything kept more or less stable, and so he
argued that surface temperatures are a function of ocean currents, affected by
the fact that the Earth's surface is like a waterbed, parts always rising or
lowering and hence changing ocean flows. Lyell, in an example seized on
by Herschel as a proper form of causal argumentation, noted that the British
Isles are much warmer than expected thanks to the Gulf Stream, a present,
physical, observable phenomenon that could then be taken as an analogy for
what is going on indefinitely. Lyell's "grand new theory of climate" (letter to
Gideon Mantell, February 15, 1830). Herschel wrote critically of those who
think the world's climate is uni-directional, from hot to cold, with the past but
an uncertain sign to the present. Nor was he much taken with the idea that
volcanoes in the past were bigger and more influential on Earth's geology than
they are today.

Neither of these is very convincing. Herschel wrote:

A cause, possessing the essential requisites of a *vera causa*, has, however,
been brought forward in the varying influence of the distribution of
land and sea over the surface of the globe: a change of such distribution,
in the lapse of ages, by the degradation of the old continents, and the

elevation of new, being a demonstrated fact; and the influence of such a change on the climates of particular regions, if not of the whole globe, being a perfectly fair conclusion, from what we know of continental, insular, and oceanic climates by actual observation. (139)

Darwin bought in entirely to Herschel's characterization of Lyell's theory as a *vera causa*. His own major contribution to geology was showing that Lyell was mistaken in thinking coral reefs are the tops of volcanoes, that just so happen barely to break the surface of the ocean, where coral then builds on the edges of the crater. Rather, argued Darwin, in 1842, the seabed is subsiding gradually and, as it does, coral keeps building up the sides of the crater, always just above the surface, where coral can survive (Fig. 5.1a, b). Another instance, where Darwin actually used the language of *vera causa*, was in 1839 after the *Beagle* voyage when he traveled up to Scotland to Glen Roy,

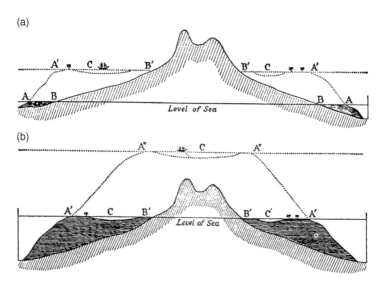

**Figure 5.1** a, The sea starts to rise and the coral, which grows only close to the surface, rises in tandem. b, The sea has now reached its present level, and all that remains of the island is a ring of coral piercing the surface.

attempting to explain the mysterious "roads" around the inside of the valley. Darwin argued, in Lyellian fashion, that there had been a lake in the valley and as the ground rose the water vanished, leaving the beaches – the roads (Fig. 5.2). As we now know, Darwin was completely wrong. The Swiss geologist Louis Agassiz showed there was an inland lake, dammed by glaciers (during the last big ice age). The glaciers melted, the lake drained out, the beaches/roads remained. There was a *vera causa*, but not Darwin's.

It is no big surprise to find that, on his route to natural selection, Darwin worked entirely in a Herschellian mode. He sought a *vera causa* for evolution, the tree of life. To recap from Chapter 1, Darwin quickly saw that artificial selection held the key. Breeders changed organisms through selection. The trick was to find out how nature could change organisms through selection. At the end of September 1838, Darwin read Malthus on population, and all fell into place. There is a huge pressure for food and space. "The final cause of all this wedging, must be to sort out proper structure, & adapt it to changes." He

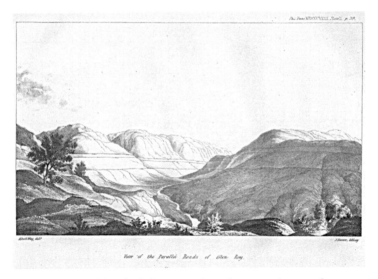

**Figure 5.2**   The parallel roads of Glen Roy: "a sheet of water must have stood at as many levels as there are buttresses."

adds: "One may say there is a force like a hundred thousand wedges trying force into every kind of adapted structure into the gaps of ~~in~~ the oeconomy of nature, or rather forming gaps by thrusting out weaker ones" (*Charles Darwin's Notebooks*, D 135e). This is the route to discovery. But as soon as he had made the connection, Darwin then used the analogy for justification. Remember: "Can the principle of selection, which we have seen is so potent in the hands of man, apply in nature? I think we shall see that it can act most effectually." And: "Can it, then, be thought improbable, seeing that variations useful to man have undoubtedly occurred, that other variations useful in some way to each being in the great and complex battle of life, should sometimes occur in the course of thousands of generations?" A Herschellian *vera causa*!

## Rationalist Consilience

What then of Whewell? He was no less Newtonian than Herschel, but whereas Herschel started with the evidence to get his *verae causae*, Whewell started with the hypotheses. In 1840, he introduced the term "consilience of inductions," meaning a theory or hypothesis that explains in different independent areas.

> Accordingly the cases in which inductions from classes of facts altogether different have thus jumped together, belong only to the best established theories which the history of science contains. And as I shall have occasion to refer to this peculiar feature in their evidence, I will take the liberty of describing it by a particular phrase; and will term it the Consilience of Inductions. (*Philosophy*, 2, 230)

This applies to Newton's theory, obviously. It explains terrestrial mechanics and celestial mechanics, two areas that under the Ptolemaic geocentric world view were kept entirely separate – the imperfect world of change versus the perfect world unchanging. Whewell had a more modern example to strengthen his case. Stemming from the seventeenth century, there were two theories of light: the wave theory (known as the "undulatory theory") of Christiaan Huygens, and the particle theory of Newton. During the eighteenth century the particle theory had prevailed; but at the beginning of the nineteenth century, facts – like those from Young's double slit experiment – started to point strongly to the wave theory. By 1830, it was all over bar the shouting.

Herschel tied himself in knots in 1827, trying to show how the theory had empirical analogies. These demanded all sorts of unlikely models involving tuning forks and sealing wax and that sort of thing. Whewell airily cut through the thickets, pointing out that the wave theory explained phenomena of different kinds:

> And of the same kind is the evidence in favour of the undulatory theory of light, when the assumption of the length of an undulation, to which we are led by the colours of thin plates, is found to be identical with that length which explains the phenomena of diffraction; or when the hypothesis of transverse vibrations, suggested by the facts of polarization, explains also the laws of double refraction. When such a convergence of two trains of induction points to the same spot, we can no longer suspect that we are wrong. (2, 447)

The conclusion follows at once: "Such an accumulation of proof really persuades us that we have to do with a vera causa."

Covering his options, as it were, Darwin bought into this too. Natural selection was not only an empiricist *vera causa*, it was also a rationalist *vera causa*. In a letter to a friendly critic, C. J. F. Bunbury, written in 1860, just after the *Origin* was published, Darwin wrote:

> With respect to Nat. Selection not being a "vera causa"; it seems to me fair in Philosophy to invent *any* hypothesis & if it explains many phenomena it comes in time to be admitted as real. In your sense the undulatory theory of the *hypothetical* ether (the undulations themselves being not recognised) is not a vera causa in accounting for all the phenomena of Light. Natural selection seems to me in so far in itself not be quite hypothetical, in as much if there be variability & a struggle for life, I cannot see how it can fail to come into play to some extent. (Letter to C. J. F. Bunbury, February 9, 1860)

## Mechanism?

To answer the question posed, Darwin did think natural selection was a cause, the cause of evolution. Moreover, it was a cause of the best kind: a *vera causa*. There were three lines of support for this conclusion. First, the analogy from artificial selection. Second, the structure of the theory, with natural selection playing the same kind of role as Newtonian gravitational attraction. Third, the consilience, running through the sub-branches of biology: social behavior, paleontology, biogeography, systematics, anatomy, and embryology. Darwin thought of natural selection as a cause. Did he then, by extension, think of natural selection as a "mechanism"? From about 1900 on, that seems to have been common talk. In 1909, Francis Darwin – Darwin's botanist son – writing the introduction to two hitherto-unpublished manuscripts by his father, said without comment: "I have no doubt that in his retrospect he felt that he had not been 'convinced that species were mutable' until he had gained a clear conception of the mechanism of natural selection, i.e. in 1838–9". Showing balance, a couple of years earlier (1907), in an already-encountered book very critical of Darwin's thinking – *Darwinism Today* by Vernon L. Kellogg – we find: "Darwinism, primarily, is a most ingenious, most plausible, and, according to one's belief, most effective or most inadequate, causo-mechanical explanation of adaptation and species-transforming."

I am not sure that we should read too much into this. It seems to be the beginning of what has become standard thinking. To take but one instance, in his book *The Selfish Gene*, without any need of explanation or excuse for his language, Richard Dawkins says that, if you start way back in time with a kind of organic primeval soup, then "if you had sampled the soup at two different times, the later sample would have contained a higher proportion of varieties with high longevity/fecundity/copying-fidelity. This is essentially what a biologist means by evolution when he is speaking of living creatures, and the mechanism is the same – natural selection" (19). However, going back to Charles Darwin himself, he seems never to have used the word "mechanism" in speaking of natural selection. He was not averse to the use of the word "mechanism." In his little book on orchids, published in 1862, again and again he referred to the "mechanisms" involved in fertilization.

> In a spike of *Orchis maculata* I found as many as ten flowers, chiefly the upper ones, which had only one pollinium removed; the other pollinium being in place, with the lip of the rostellum well closed up, and all the mechanism perfect for its subsequent removal by some insect. (*On the Various Contrivances*, 19)

Moths had removed the pollinia, and had thoroughly well fertilised the perfect flowers on the same spike; so that they must have neglected the monstrous flowers, or, if visiting them, the derangement in the complex mechanism had hindered the removement of the pollinia, and prevented their fertilisation. (47–8)

When the pollinia are exposed to the air the caudicle is depressed in from 30 to 60 seconds; and as its anterior surface is slightly hollowed out, it closely clasps the upper membranous surface of the disc. The mechanism of this movement will be described in the last chapter. (80)

The use of the term "mechanism" is restricted to a machine-like contraption, such as the mechanism of the pump that is used to raise water to a higher level, or the mechanism of a watch for keeping exact time. In other words, it is always a metaphor for something devised by humans. As expected, the big influence here was natural theology – the argument from design – showing for instance that the eye (made by God) is analogous to the telescope (made by humans). In the orchids book, Darwin made explicit reference to Sir Charles Bell's contribution – *Mechanism and Vital Endowments of the Hand as Evincing Design* – to a celebrated series of works of natural theology: the "Bridgewater Treatises." Before this was Paley's *Natural Theology*, where the machine metaphor is working flat out. Take the argument that the eye is like a telescope: "To some it may appear a difference sufficient to destroy all similitude between the eye and the telescope, that the one is a perceiving organ, the other an unperceiving instrument. The fact is that they are both instruments. And as to the mechanism, at least as to mechanism being employed, and even as to the kind of it, this circumstance varies not the analogy at all." Obviously, natural selection is not a mechanism in this sense, and – speaking now from the horse's mouth, as one might say – one doubts that any of the later writers who spoke of natural selection as a mechanism thought of it as a mechanism in this sense. It has become just an alternative word for cause.

## Did Darwin Make His Case?

Pick up again on the putative causal nature of natural selection. Darwin thought it was a cause. Was he right? His closest supporters were far from sure. Thomas Henry Huxley always thought that the Herschellian *vera causa* case had not been made.

That this most ingenious hypothesis enables us to give a reason for many apparent anomalies in the distribution of living beings in time and space, and that it is not contradicted by the main phenomena of life and organisation appear to us to be unquestionable; and, so far, it must be admitted to have an immense advantage over any of its predecessors. But it is quite another matter to affirm absolutely either the truth or falsehood of Mr. Darwin's views at the present stage of the inquiry … The combined investigations of another twenty years may, perhaps, enable naturalists to say whether the modifying causes and the selective power, which Mr. Darwin has satisfactorily shown to exist in Nature, are competent to produce all the effects he ascribes to them; or whether, on the other hand, he has been led to over-estimate the value of the principle of natural selection, as greatly as Lamarck over-estimated his *vera causa* of modification by exercise.

Huxley was particularly concerned that no one yet had been able to show that selection could lead to speciation – intergroup sterility. A bulldog and a greyhound look very different, as much as the members of different genera. Yet they can still produce fertile hybrids. A major problem here was that no one was much inclined to set out to show Huxley wrong. Darwin himself, as we have seen, was not ready or prepared to set up experiments to show how selection could bring on speciation, backing this by looking in nature at cases where speciation was clearly on its way. Island populations, perhaps? Huxley simply had no interest in showing that selection could in fact cause speciation. As we have seen, selection and consequent adaptation was, if anything, a hindrance to him a morphologist. All he really wanted was evolution to use as a kind of metaphysical background, and this more for general debate – arguing publicly with the Bishop of Oxford – than as something to use in his science.

Today, as you might expect, things have much changed. We have evidence of speciation through selection, both from experiments and from studies in nature. A classic, human-driven experiment involved corn (maize). Two varieties, white and yellow, were planted intermingled in a field. Each year, selection was applied, favoring only those specimens that showed least inclination to hybridize. In a mere five years, the intercrossings of the yellow variety declined from 46.7 percent to 3.4 percent, and of the white variety, from

35.8 percent to 4.9 percent. The proximate causes were no secret. The white variety started to flower at an earlier time and the yellow variety at a later time. It is easy to suppose natural conditions where something similar might obtain, although it should be noted that this was a simulated case of "sympatric" speciation, where no geographical isolation is involved. To be fair, there are evolutionists – Ernst Mayr, most vocally – who doubt that this could ever occur naturally; but, to assuage worries, there are many experiments (especially with fruit flies) showing that "allopatric" speciation, where populations are geographically separated, can happen. These involve such features as temperature and humidity tolerance, adaptation to DDT, and photo- and geotaxis (attraction to light and gravity).

An equally classic natural experiment involves "ring-species." You sometimes find a chain of subspecies, each touching and interfertile with its neighbors, but at the ends, if the two subspecies meet, they are reproductively isolated: speciation caught in action. The classic example is the greenish-warbler complex (*Phylloscopus trochiloides*) in Asia. You have a ring of subspecies (actually, today, rather broken because of deforestation), and molecular evidence points to close relationship between neighbors and some interbreeding. However, the end (touching) subspecies do not interbreed – they have different songs and there is no recognition between members of the two groups. They are different species (Fig. 5.3).

What then of the other criteria of *vera causa* status? It was Darwin's great American supporter, Asa Gray, who was the sceptic here and who wrote in 1876:

> From the very nature of the case, substantive proof of specific creation is not attainable; but that of derivation or transmutation of species may be. He who affirms the latter view is bound to do one or both of two things: 1. Either to assign real and adequate causes, the natural or necessary result of which must be to produce the present diversity of species and their actual relations; or, 2. To show the general conformity of the whole body of facts to such assumption, and also to adduce instances explicable by it and inexplicable by the received view, so perhaps winning our assent to the doctrine, through its competency to harmonize all the facts, even though the cause of the assumed variation remain as occult as that of the transformation of tadpoles into frogs, or that of Coryne into Sarzia.

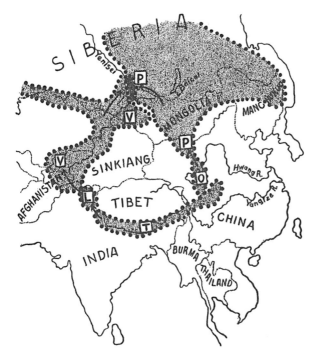

**Figure 5.3** Ring species. The ends of the ring touch at the top left (hatching).

The first line of proof, successfully carried out, would establish derivation as a true physical theory; the second, as a sufficient hypothesis.

These two demands are for proof of the adequacy of Darwin's second claim, about selection being part of a successful hypothetico-deductive system, and about the adequacy of his third claim, about selection fitting the Whewellian rationalist *vera causa* criterion. Taking the third claim about the consilience, Gray stated bluntly that we have suppositions and not proofs. "An opponent might plausibly, and perhaps quite fairly, urge that the links in the chain of argument are weakest just where the greatest stress falls upon them." It was here that, as we have seen, Darwin retorted that he had done the necessary. "With respect to Nat. Selection not being a "vera causa"; it seems to me fair in Philosophy to invent any hypothesis & if it explains many phenomena it comes in time to be admitted as real."

Was he right? This is still a much-contested topic. Here, let us say simply that Darwin has been supported in many areas of science since the *Origin* was published. One thinks of biogeography and the revolution caused by the coming of plate tectonics and continental drift, just over half a century ago. It is hard to see how any of it would make any sense without a background of evolution – an evolution driven by a slow, natural process like selection. More concretely, perhaps, in the field of social behavior, formal models have been devised giving support to Darwin's hypotheses – speculations, if you like – about behavior and relatedness in social animals, particularly in the hymenoptera. More on this in the next chapter.

## Is Natural Selection Improperly Prioritized?

What about the middle criterion? Recall, Darwin had written that "In fact the belief in natural selection must at present be grounded entirely on general considerations. (1) on its being a vera causa, from the struggle for existence; & the certain geological fact that species do somehow change." Here, again, Gray demurred. We do not have "real and adequate causes." At least part of the problem here was that Darwin did not offer anything like a formal system, as did Newton, and so at best we have a lot of "possibilities" and "maybes" and "it cannot be denied thats" and so forth. Also, Darwin did not give us selection as a cause in an initial premise, but derived it as such along the way, from the struggle and from variation. I am not sure how much of a problem this is, although it is certainly true that it leaves much to be said about variation.

Here it is worth making the point that, as we have seen in the previous chapter, variation is the one area where Darwin's theory has been supplemented significantly with the advent of theories of heredity, first Mendelian and then molecular. In major respects, the theory has been made more Newtonian. Newton's first law of motion is an equilibrium law. If nothing happens, nothing happens. "An object remains at rest (if originally at rest) or moves in a straight line with a constant velocity if the net force on it is zero." You can now introduce gravity to mess things up, knowing that, if and when things do get messed up, it must be gravity at work because things do not mess up on their own. As we have seen, today's neo-Mendelian version of Darwin's selection theory starts with its own equilibrium law, the so-called Hardy–Weinberg law. In an infinitely large population, with no external forces impinging, gene ratios stay the same.

In a serious way, one is starting to speak to the issues that worried Asa Gray. But does this backfire? You are now agreeing with Sewall Wright, for instance, that selection is just one of a number of factors behind evolution, which is now seen as something that does not start with natural selection, rather with the Hardy–Weinberg equilibrium, which now can be upended by a number of factors. Selection, certainly, but also mutation, immigration in and out of the group, and plausibly genetic drift sometimes. There seems to have been an awful lot of energy going into the *vera causa* issue, when it seems that we are talking about but one part of whole. Philosopher Elliott Sober makes much of this. In particular, he highlights "ancestral influence" – how the organism's past affects its present and future – and asks "whether it is worth considering whether Darwin should simply have said that selection has been a *very important* cause of evolution. Was it needlessly audacious to put selection at the top of a list [of putative causes behind evolution] whose members he had no reason to think he could completely enumerate?" (*Did Darwin Write the Origin Backwards?* 21). He answers his own question: "is 'evolution by natural selection' a good characterization of Darwin's theory? The answer is emphatically *no*." Well, that tells it like it is!

Sober is not alone in making this kind of charge. Philosopher Richard G. Delisle has recently offered a similar kind of downgrading. He goes so far as to say that natural selection is a "mere auxiliary hypothesis." He tells us that in the *Origin*, "pattern comes first with process being merely superimposed upon a pre-established pattern." The central element in Darwin's theory is the tree of life, and all else is clean-up. "Darwin put natural selection in a conceptual strait-jacket, depriving it of its evolutionary creativity and reducing it to an auxiliary hypothesis." Among the really important factors of change, bringing about the tree, selection does not really figure. Far more important, indeed the most important in Delisle's treatment, is "the principle of divergence, whose function is to ensure that extinct forms fall in-between extant ones (thus preventing the rise of variations unrecorded among extant forms)" (*Natural Selection*, 73). The picture seems to be a kind of growth in an organic manner, with divergence then making for branching and ultimately the different forms. "For Darwin, not only does 'pattern' always comes first, but it is also envisioned in such a rigid way as to leave no room for evolutionary change other than 'divergence'" (73). One really starts to feel sorry for natural selection: "the anticipated relationship between process (natural selection) and pattern (divergence)—the action of natural selection instigating divergence—is shattered and replaced by the inherent connectiveness of life, whose natural pattern imposes itself upon a natural selection that is powerless at undoing it" (78). There is more, much more in this mode, but you start to get the idea.

Doubling back to Sober, whom Delisle quotes as having got the general picture "nicely," there is no big surprise that Sober like Delisle makes the tree of life central. This indeed is flagged by his title "Did Darwin write the *Origin* Backwards?" Sober's position is that the history – the tree of life – is prior to any causal explanations, hence the title. Indeed, Sober goes so far as to suggest that this ordering was really Darwin's also. Seizing on Darwin's insight that we have already documented, a function of his work on barnacles, Sober stresses that it is homology that counts in these cases – non-adaptive homology. He spells out what he calls "Darwin's Principle."

> Adaptive similarities provide almost no evidence for common ancestry while similarities that are useless or deleterious provide strong evidence for common ancestry. (25)

I suspect that Darwin would be somewhat puzzled – upset – at the implication that this is his major theoretical contribution to evolution studies. Apart from anything else, we have seen that he firmly puts Unity of Type (homology) behind and a consequence of Conditions of Existence (adaptation). To find out what has gone wrong, it is instructive to recall that Darwin's model for his theory was Newtonian mechanics. In that case, it was certainly true that the finding of the phenomena – planets circulating the Sun (Copernicus), planets going in ellipses (Kepler), cannon balls going in parabolas (Galileo) – was temporally prior to the finding of Newtonian causes. Kepler, remember, was much in the organicist paradigm as opposed to the mechanist paradigm within which Newton worked. The last thing he was after was the force of gravitational attraction. However, obviously, Newton's ideas were conceptually prior to those of the earlier physicists. The same is true in the Darwinian case. The tree of life is temporally prior to the causes. Remember, Darwin came up with the tree of life early in 1837 (Fig. 1.3), but it was late in 1838 that he found natural selection. But this does not mean that, in the *Origin*, Darwin was wrong to make natural selection conceptually prior – the very opposite, in fact. For this reason also, there is no great surprise that working out the tree of life is not primarily a job for natural selection, any more than working out the elliptical paths of the planets is a job for gravitational attraction. Indeed, as we saw earlier, the people who today are taken as having the best way of finding phylogenies, the cladists, are almost contemptuous of natural selection. As David Hull wisely noted, they could not care less about causes. This is no great surprise because the father of cladism, Willi Hennig, was a German biologist, firmly on the organicist side of things. For him, for the cladists, basically for anyone working out paths – and this as we have seen includes Darwin – it is homology that counts. Adaptation is a nuisance.

So truly, while what Sober says about working out phylogenies (ancestries) is true, it has no bearing on the status of causes. We have seen reason to think that Darwin was right to prioritize natural selection – he never thought that other forces like the Lamarckian inheritance of acquired characteristics were dominant, and the same thinking about other causes (for instance, drift) is true today. Furthermore, to finish this discussion, Delisle's prioritizing of divergence backfires badly on his position. As we have seen, divergence is a function of the division of labor, and that is entirely selection-driven. Remember: "the more diversified the descendants from any one species become in structure, constitution, and habits, by so much will they be better enabled to seize on many and widely diversified places in the polity of nature, and so be enabled to increase in numbers."

Of course, this is not to say that selection necessarily does everything. Darwin's position about reproductive barriers between species was that they are a function of different reproductive systems being unable to coordinate one producing hybrid offspring. In an 1868 response to Wallace, who – arguing that the sterility of the mule could be of advantage to the parent species, horse and donkey – suggested that selection could be of groups, in particular of species, Darwin responded unequivocally.

Let me first say that no man could have more earnestly wished for the success of N. selection in regard to sterility, than I did; & when I considered a general statement, (as in your last note) I always felt sure it could be worked out, but always failed in detail. The cause being as I believe, that natural selection cannot effect what is not good for the individual, including in this term a social community.

There are no extreme claims that everything about the tree of life can be explained directly by natural selection. Parallel instances of evolution are surely often a function of one way of doing (or changing) things is better physiologically than alternative ways– so different lines do the same thing. But whoever said that selection always does everything? The white coats of polar bears are as much a function of cold weather, snow and ice as they are of natural selection.

In short, once again Darwin made his case for the *vera causa* principle and its applicability to selection. It is not, and should not be considered as, a second-class cousin.

# 6 The Positive Case

A little arbitrarily, but not entirely without reason, let us take 1959, the 100th anniversary of the *Origin*, as the date when the Darwinian paradigm finally came into its own. Natural selection and Mendelian genetics, now rapidly becoming molecular genetics, gave the explanation of the tree of life. If we continue to think in Kuhnian terms, what now of normal science? We should expect to see the subbranches of the consilience come into their own, as practitioners moved forward, theoretically, experimentally, and in nature, raising and solving their problems. And in major respects we do see exactly this.

## The Consilience

Biogeography was transformed by the new theory of plate tectonics, and neo-Darwinism – as we might call the new paradigm to stress the selection element – moved in as a co-partner, helping to find and understand the distributions of the organisms around the globe. More fine-grained work was also possible. An early classic was *The Theory of Island Biogeography*, by Robert H. MacArthur and Edward O. Wilson. They showed how the numbers of species on islands are a (selection-driven) function of the size of the islands (increase) and the distance from the mainland (decrease), offering a theoretical analysis including predictions about equilibria (Fig. 6.1). They then went on to discuss how various factors would affect outcomes. One highly influential idea was that, on islands – later generalized – there will be an evolution from species having many offspring with little parental care ("r-selection") to species having few offspring with much parental care ("K-selection").

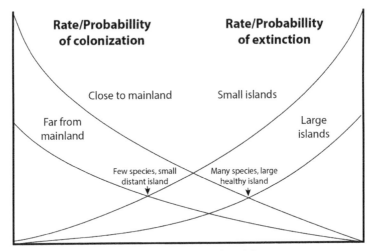

**Figure 6.1** Island biogeography.

Paleontology was another area opening up new discussions and hypotheses. Best known, perhaps, was the theory of "punctuated equilibria" promoted by Niles Eldredge and Stephen Jay Gould in 1972. Their claim was that jumps in the fossil record from one form to another do not necessarily represent incomplete evidence (with fossils of linking forms no longer in existence), but rather evidence of rapid evolution from one form to another. Contrary to popular opinion – although one can certainly read this in a non-Darwinian fashion as organisms simply changing in one generation to another form (saltationism) – this theory does not necessarily deny Darwinian forces. The theoretical claim is that small populations get isolated. Since all larger populations have, as we have seen, a considerable amount of (genetic) variation, any small population is going to be atypical compared to the parent population. Some variations will be in the small population. Some will not. This gives rise to the "founder principle" – the new, small population will undergo rapid evolution (too quick to be reflected in the fossil record), as its limited variations settle down to function together. The newly isolated populations will thus settle into viable new species. Rapid change is not only this result of the genes settling into functioning efficiently together, but combined probably with the effects of new environmental circumstances for the isolated group. Lots of work for selection here.

Another area where paleontology has seen advances – a sub-area usually known (for obvious reasons) as "paleobiology" – is through the use of computers processing large amounts of data. John J. Sepkoski Jr. studied the evolution of marine animals since the beginning of the Cambrian (570 million years ago). He found systematic patterns, as what one might call waves of new kinds entered the picture. Using ideas taken from the MacArthur–Wilson model of biogeography, he suggested that initially the forms are fairly generalized and can occupy unused niches rapidly, but then the niches become full, and evolution slows down in the direction of ever-more specialized forms. Then a new form appears, and the story starts over again. The accompanying figure (Fig. 6.2) shows the three waves of marine life

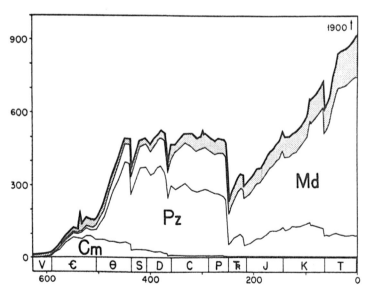

**Figure 6.2**   Sepkoski species diversity – or mystery of mysteries.

in the past 600 million years: first less sophisticated animals like trilobites, then starfish and more, and finally complex animals like reptiles – from primitive to advanced. "From an evolutionary standpoint, it is hardly surprising that a more advanced fauna would succeed a more primitive one in time." (There is no discussion of how this progress might occur. What perhaps Sepkoski is pre-supposing is that things overall tend to complexity. We shall look later at this issue.)

## Micro- Versus Macro-evolution

This is a selection-based history of life's changes. First one kind of form – unspecialized – has the advantage, then as things change and niches get filled up, other kinds of forms – specialized – take over. Is this jumping the gun rather? Let us agree that selection can be effective in relatively small issues – color of a butterflies' wings for example. Microevolution, as one might say. Can one just extend this up to major changes over time – macroevolution? There are those who doubt it. Historian Mark Adams wrote: "Darwin's book title had emphasized the origin of species, and he apparently believed that if natural selection could explain that, it could also explain the origin of genera, families, orders, and other higher taxa". Adams argued that this "has remained contested territory, and biologists have debated over exactly where the divid-ing line between 'microevolution' and 'macroevolution' should be drawn. Consistent throughout, however, has been the core problem of Darwin's 'natural selection' theory, namely, its questionable relevance to what most scientists and the lay public have meant by 'evolution.'" In the end, argued Adams, neo-Darwinians (and he included here American synthetic theorists) have rather fudged the issue, simply assuming that one can go from one to the other. The paleontologist G. G. Simpson, for instance, admitted the gap in 1944.

> Micro-evolution involves mainly changes within potentially continuous populations, and there is little doubt that its materials are those revealed by genetic experimentation. Macro-evolution involves the rise and diver-gence of discontinuous groups, and it is still debatable whether it differs in kind or only in degree from micro-evolution. If the two proved to be basically different, the innumerable studies of micro-evolution would

become relatively unimportant and would have minor value in the study of evolution as a whole. (*Tempo*, 97)

In the end, surmised Adams, it was all more a matter of prior commitments, perhaps a sociological matter of science versus religion, than an appeal to firm scientific reasons. In the United States, Creationists were going after evolution. "This situation may well have encouraged individual scientists to downplay their reservations about the macroevolutionary question in order to protect evolutionary biology as a whole: uncertainties about mechanism may have seemed less pressing when the validity of evolution itself was under attack, especially when those uncertainties were being deployed by the creationists to disprove evolution" (*Evolution*, 217).

One doubts that many professional evolutionists today would agree with this assessment. In an optimistic overview of the state of play in selection studies today – "From low to high gear: there has been a paradigm shift in our understanding of evolution" – David Reznick and co-workers set up the challenge: "The divide between microevolution and macroevolution is most often recognised by a distinction between genetic changes within populations and speciation". They respond that today, regular selection-fueled evolution can bridge the gap. Rosemary and Peter Grant, in their long-scale studies of Darwin's finches on the Galapagos, have shown that, in short pieces of time, you can get changes that any reasonable person would call "macro". And indeed, this example does not stand alone. There are many other examples, including sticklebacks, mosquitoes in the London Underground, and host races of the fruit fly *Rhagoletis pomonella*.

In the next chapter, we pick up the question of whether changes like these truly qualify as "macroevolutionary."

## Social Behavior

Hovering behind much of the discussion is what is known as the "levels of selection" problem. What does natural selection select? Darwin had no doubt. It selects for individual organisms. Remember the response to Wallace, disagreeing flatly with the suggestion that selection could work for the benefit of species. He did not buy into the argument that the sterility of the mule could be of advantage to the parent species, horse and donkey. Rather than go down that road – group selection – he accepted that the sterility was a by-product of different reproductive systems.

Today, as we have seen, evolutionists are happy to extend the notion of individual from the organism to the gene. Shortly, we shall look at the connection between the two, but for now, using the useful terminology of David Hull, we can speak of the genes as the "replicators" – the units that go on from generation to generation – and the organisms as the "interactors" – the units that strive in the struggle for existence. Are there other higher levels of selection? Sepkoski seemed to be talking of some kind of "species selection," where species of a certain kind – in certain circumstances specialists – do better than species of another kind – generalists. I doubt that most people would take umbrage at this – the members of one kind of species are fitter than the members of another kind of species. But notice we are offering more of a tabulation than a causal explanation. There is no question of two species competing against each other, or at least there is no reference to that here. It is not the species that are being selected, but the individual organisms, which can then be categorized.

More controversially, is there place for something in between, for a kind of group selection, where being a member of a group is the decisive issue – having features that only make sense in the context of a group? Here Darwinian selectionists draw the line, agreeing with Darwin. Why? Because of the problem of cheating. If it is the group (as opposed to the individual) that is succeeding in the struggle for existence, then that means that some members are (actually or potentially) giving to the group even though they get nothing in return. They are pure altruists. The trouble is that, in such a group, those who are not altruistic are at an advantage to those who are. The non-altruists can rely on the efforts of themselves and of the altruists, whereas the altruists lose out because the non-altruists are not going to help them. They will be less fit and eliminated, and group selection comes to an end.

Is this all that there is to be said? What about Darwin's "social communities"? As we saw in Chapter 1, despite little understanding of the nature of heredity, Darwin did see that relatedness could explain adaptations in organisms, even if they themselves did not breed. It could be the success of the relatives that counts. Family selection! Darwin wrote in the *Origin*:

> I have such faith in the powers of selection, that I do not doubt that a breed of cattle, always yielding oxen with extraordinarily long horns, could be slowly formed by carefully watching which individual bulls and cows, when matched, produced oxen with the longest horns; and yet no one ox could ever have propagated its kind.

Remember: "Thus I believe it has been with social insects." If something in the sterile offspring helps the community at large, then selection will direct the fertile members of the group to keep producing such advantageous sterile offspring. "And I believe that this process has been repeated, until that prodigious amount of difference between the fertile and sterile females of the same species has been produced, which we see in many social insects" (238).

This kind of family selection is now known as "kin selection," and Darwin was on solid ground here. Building on the insights of William Hamilton in 1964, we have considerable understanding of how things do work – understanding that can be quantified and that has been successfully tested. Like so many great ideas, "Hamilton's rule" is straightforward – for kin selection to kick in, the product of the relatedness of an organism doing the action, $r$, and the benefit to the recipient, $B$, must be greater than the cost of the action to the organism giving help, $C$. In other words: $rB > C$.

Using the term "altruism" to mean giving and expecting benefits (rather than disinterested altruism where benefits are neither expected nor do they exist), then how does such altruism kick in? Hamilton wrote in 1964:

> The selective advantage which makes behaviour conditional in the right sense on the discrimination of factors which correlate with the relationship of the individual concerned is therefore obvious. It may be, for instance, that in respect of a certain social action performed towards neighbours indiscriminately, an individual is only just breaking even in terms of inclusive fitness. If he could learn to recognise those of his neighbours who really were close relatives and could devote his beneficial actions to them alone an advantage to inclusive fitness would at once appear. Thus a mutation causing such discriminatory behaviour itself

benefits inclusive fitness and would be selected. In fact, the individual may not need to perform any discrimination so sophisticated as we suggest here; a difference in the generosity of his behaviour according to whether the situations evoking it were encountered near to, or far from, his own home might occasion an advantage of a similar kind. (*Genetical evolution*, 51)

"Personal fitness" is the benefit directly to an organism because of its behavior; "inclusive fitness" is the indirect benefit to an organism because of its behavior. Thus, if an organism has two offspring, its personal fitness is $\frac{1}{2} + \frac{1}{2} = 1$. If an organism has four nieces or nephews, its inclusive fitness is $\frac{1}{4} + \frac{1}{4} + \frac{1}{4} + \frac{1}{4} = 1$. Eight cousins, 1/8 related, is the same. J. B. S. Haldane quipped: "I would lay down my life for two brothers or eight cousins."

## Human Evolution

As is well known, as is only too well known, new models of social behavior like that of Hamilton led to a rapid development and expansion of work on social behavior. It was not long before it was being applied to that most tempting of organisms, *Homo sapiens*, most famously or notoriously by Edward O. Wilson, of island biogeography fame, who in a wide-ranging survey right up to our own species – *Sociobiology: The New Synthesis*, published in 1975 – argued that we now have the key to unlocking many secrets about animal actions, for instance, parental behavior, male/female roles and relationships, same-sex behavior, and much more. And at once, criticism flared up, perhaps expectedly from social scientists who saw their domain entered by would-be all-conquering biologists, and less expectedly from Wilson's fellow biologists – often Marxists – who argued that his thinking was redolent of the biological deterministic thinking of the worst excesses of the 1930s. Looking back, it is not immediately obvious that much advance was made on the biological front, although the controversy did indeed provide grist for the mills of many nascent philosophers of science writing their dissertations. Given new names, "evolutionary psychology" for instance, work continued happily on. Rather than attempt an overview – in the next chapter we shall again raise the critique – we end this chapter with a brief look at the work of paleoanthropologists and paleoarcheologists who have started applying Darwinian principles, chiefly selection, to the problems that arise in their domain.

Focus on war. If anything is a widespread human phenomenon it is war. In the last century there was the First World War, twenty to forty million dead; the Second World War, sixty to eighty million dead; the Russian Civil War, five to ten million dead; the Chinese Civil War (1927–49), about ten million dead. And that is just a start, via Korea and Vietnam, down to Iraq and Afghanistan. Is war part of human nature? Many would argue that it is. We are, in the popular lingo, "naked apes." As anthropologist Raymond Dart put it in 1953:

> The blood-bespattered, slaughter-gutted archives of human history from the earliest Egyptian and Sumerian records to the most recent atrocities of the Second World War accord with early universal cannibalism, with animal and human sacrificial practices of their substitutes in formalized religions and with the world-wide scalping, head-hunting, body-mutilating and necrophiliac practices of mankind in proclaiming this common bloodlust differentiator, this predaceous habit, this mark of Cain that separates man dietetically from his anthropoidal relatives and allies him rather with the deadliest of Carnivora.

## A Darwinian Approach

Against this, the Darwinians suggest that we might first look to natural selection and learn what it tells us. The anthropologists and archeologists tell us that humans broke from the apes rather more than five million years ago: they left the jungles, went out on the plain, became bipedal and started their long journey up from brains about 400 cc (chimpanzee size) to 1,200 cc (human size). For almost all the time, we were hunter-gatherers in bands of about fifty, usually fairly isolated, in a vast, vast territory (Fig. 6.3). Ask now about the selective value of being killer apes. It isn't high. The route we took was that of sociality. In a small band, you need to get on with others. United we stand, divided we fall. Our adaptations drove us that way: not just the brains that enabled us to work things out, within and without the group – where are the best hunting patches, for example – but also anatomical adaptations, for instance in losing those attributes (like big teeth and claws) that bring on dangerous violence, as well as physiological, for instance in our females not

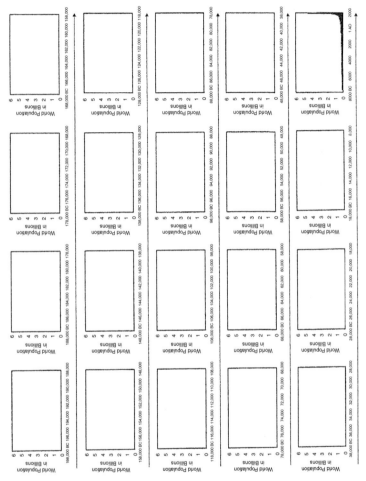

**Figure 6.3**   A pictorial representation of the vastness of the world in which hunter-gatherers lived. For almost all the time of their existence, they were so few, and the world so large, that there was no good Darwinian reason to go to war with each other. This occurred when, as shown in the last rectangle, population numbers started to explode, agriculture started, people then had assets worth fighting for, and the inability simply to move away out of danger. (Rotate the figure 90 degrees clockwise.)

coming into heat and causing a down-tools for a day or two, about every other week. We worked together within the group.

How then were we going to regard the members of other groups? One thing you can guarantee is that we were not about to attack them on sight. We might get hurt! There certainly might be a reason for a degree of wariness – we wouldn't want them pinching all our fertile women nor do we want them seducing our young braves into the thought that the grass is greener on the other side. These sorts of things are going to happen – and in the face of certain disasters like Ice Ages there might well be reasons to work together – but, generally, the best policy, in times of tension, is to get up and go away.

Then came agriculture. This had two immediate consequences. First, the population numbers soared. Hunter-gatherer bands must have a tight control on reproduction. No band can afford to have half its members under ten. Infanticide had to be a common practice. With agriculture, it is all hands to the job, and children can do many of the simpler tasks. As Daniel Lieberman put it: "farmers pump out babies much faster than hunter-gatherers" (*Human Body*, 188). Second, groups now had fixed assets. They could not simply strike camp and move on. With a population explosion, there is going to be pressure for more land, more goods. The consequence? War appeared. It didn't just happen.

Not just war, but increased prejudice against and oppression of members of out-groups, starting with women. In hunter-gatherer groups, women must play an equal role – if not hunting then gathering, if not dragging the prey back home then building traps to catch smaller animals. In bands making their way, always facing challenges, you cannot afford to have weak sisters – using this offensive, metaphorical term in an entirely appropriate, literal sense. Now, with the coming of agriculture, they became baby machines, with much reduced time for anything else. Males took over. Similar sorts of stories can be told about other out-groups. Foreigners – Brexit and Trump. If you are a hunter-gatherer and you don't much like them, then you can move away. If you are a farmer, you are stuck with them. Tensions build. Natural selection made us the way we are, and then came cultural changes for which our adaptations are not adequate. In the words of Leda Cosmides and John

Tooby: "Our modern skulls house a stone age mind." They expound: "In many cases our brains are better at solving the kinds of problems our ancestors faced on the African savannahs than they are at solving the more familiar tasks we face in a college classroom or modern city" (11).

## Grounds for Optimism?

Let us not end this chapter on a gloomy note. Bolstering the case for the importance of a general knowledge of natural selection and its workings, let us point out that our knowledge of how we got where we are is also the guide to improvement. This does not imply that we need to get involved, giving natural selection a hand and altering us genetically. We have adaptations leading us away from war and prejudice. We already have the needed genes. It is now a question of manipulating our culture so that the unpleasant effects of modern life can be minimized, if not avoided. In the case of war, for instance, institutions like the United Nations are far from perfect, but they are a good start. Every account of the beginning of the First World War stresses how people bumbled into it. The Kaiser, for instance, was totally inadequate as a leader, constantly changing his mind, refusing to take advice, ever insecure – a withered arm did not help – and trying to prove his masculinity. The Tsar was little better. And the leaders of the Austro-Hungarian empire wanted no more than to smash the Serbians and thus show who was in charge. That this would at once start the Serbians' allies, the Russians, to mobilize was simply not taken into account. At the very least, an organization like the United Nations might have got everyone to sit around a table and talk about things.

Likewise with prejudice. If natural selection did not make women inferior – if natural selection in fact did no such thing, making them (when hunter-gatherers) equals – then reversals (from states brought on by the coming of agriculture) are surely within the realm of the possible. As, indeed, the last hundred years have shown. Labor-saving devices like washing machines have freed women from many hours of drudgery. Efficient contraception has enabled women to control family numbers. No longer are they faced with the burden of raising ten children. Things are far from perfect, and the Luddites opposing abortion show that there are miles to go, but the journey has been started. As developmental psychologists Stephen Ceci and Wendy Williams explain:

Roughly half the [American] population is female, and by most measures they are faring well academically. Consider that by age 25, over one-third of women have completed college (versus 29% of males); women out-perform men in nearly all high school and college courses, including mathematics; women now comprise 48% of all college math majors; and women enter graduate and professional schools in numbers equal to most, but not all fields (currently women comprise 50% of MDs, 75% of veterinary medicine doctorates, 48% of life science PhDs, and 68% of psychology PhDs). (5)

Natural selection is a very powerful tool for understanding; morally important, as well as an important addition to the realm of pure knowledge. If it is discarded, then you suggest an alternative.

# 7 Time for a Change?

Turn now to those who think natural selection is vastly overrated as a cause of evolutionary change. It is at best a clean-up process after the real creative work has been done. It is little surprise that these critics come from within the organismic model, implicitly or explicitly. At the scientific level, we have encountered already the most (and properly) distinguished of them all, the American population geneticist Sewall Wright. Remember his "shifting balance theory," where the key lay in genetic drift, as gene levels fluctuated randomly in small subpopulations, and then, when new adaptive features appeared, the subpopulations rejoined the larger group (probably the species), and through a form of group selection the new feature spread through the whole group. This is highly Spencerian – infused with a solid dose of Bergsonian vitalism – as equilibrium is disturbed and then regained at a higher level, part of an overall progressive process, presumably ending in humankind.

## Organicist Science

Wright was working at the time when the English-transplant at Harvard, Alfred North Whitehead, was expounding his "process philosophy" – a system entirely within the organicist paradigm, with deep debts to Schelling. Explicitly, Whitehead advocated in 1926 "the abandonment of the traditional scientific materialism, and the substitution of an alternative doctrine of organism" (*Modern World*, 99). There were those who took up this thinking and applied it to religious understanding, arguing that in some sense God Himself is evolving. It is hardly surprising that this way of thinking, "process theology," is highly controversial. Fortunately, it is not our worry here. In the secular world, the British biologist Conrad Waddington was ever an enthusiast. He had a theory of "genetic assimilation," where he thought he could explain Lamarckian effects through conventional genetical theory. He subjected fruit flies to heat-shocks, which led to the hatching of flies with non-standard wings. (The wings of flies have radiating veins. There are shorter veins – "cross-veins" – linking the main veins. See Fig. 7.1.) In a few generations, through judicious selection, he could get flies with non-standard wings (wings without the cross-veins) in the absence of heat-shocks. That this kind of result showed that all led progressively up to humankind was a given. As he wrote in 1960, "The capacity to remain relatively independent of the environment, to incorporate into the life-system more complex functions of environmental variables, and ultimately to control the environment, of course reach a much higher point in man than in any pre-human species" (*Ethical Animal*, 37).

A third example of a non-Darwinian – meaning non-selection-based – approach came in 1979, in a very much discussed article in the *Proceedings of the Royal Society* – "The spandrels of San Marco and the Panglossian paradigm: a critique of the adaptationist programme" – by paleontologist Stephen Jay Gould and geneticist Richard Lewontin. In Gould's earlier writings, he made heavy use of what we have seen called the "founder principle," the claim that small, isolated populations (say, immigrants to an island) are going to be statistically different from their parent populations, because the variation in the parent populations means that the small populations will

Normal phenotype            Cross-veinless phenotype

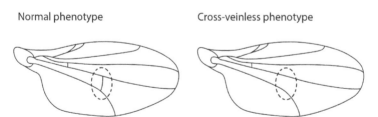

**Figure 7.1**    The wing on the left shows how the development occurs in normal circumstances. There is a gene that produces cross-veins. After heat-shock, in about forty percent of cases, as shown by the wing on the right, one gets no cross-veins.

almost certainly have more of some alleles and fewer of others. There is obviously a random factor here – a flock of birds blown out to sea in a hurricane will, by chance, be statistically different from their parent population. There is also (as noted earlier) selection, both in creating the variation in the parent population and, almost certainly, in the newly isolated flock, drawing on a depleted gene-library in a new environment, which is going to have hitherto-unfaced adaptive demands – new predators and so forth. Because of the intensity of the selective forces, one expects rapid change – change too fast to be recorded in the fossil record. By the end of the decade, Gould was arguing in a different, non-selection-friendly mode. At least part of the motivation for the shift was that Gould was ardently promoting the essential role of paleontology in the Darwinian synthesis. Although the general public usually thinks first of the fossil record when they think of evolution – not for nothing did the leading Creationist Duane T. Gish title a best-selling book *Evolution: The Fossils Say No!* – Gould felt, with some reason, that geneticists and others working on causes tend to downplay the physical evidence of the past. It is to be explained rather than to do the explaining. Gould was determined to raise the status of his discipline – to claim a seat at the "high table" (referring to the privileged place of faculty in the dining halls of Oxford and Cambridge colleges) – and to this end he was speculating in non-Darwinian jumps, saltations, as the essential behind his vision of evolution: punctuated equilibrium.

This was hardly going to appeal to the geneticist Lewontin; but, remember, this was the height of the sociobiology controversy, with Lewontin uniting with Gould in the attack. To downgrade the credentials of Wilson's Darwinian science, Gould and Lewontin argued that natural selection is not the main force for change.

An adaptationist programme has dominated evolutionary thought in England and the United States during the past 40 years. It is based on faith in the power of natural selection as an optimizing agent. It proceeds by breaking an organism into unitary 'traits' and proposing an adaptive story for each considered separately. Trade-offs among competing selective demands exert the only brake upon perfection; non-optimality is thereby rendered as a result of adaptation as well. We criticize this approach and attempt to reassert a competing notion (long popular in continental Europe) that organisms must be analysed as integrated wholes, with Baupläne [a modern equivalent of Owen's archetypes] so constrained by phyletic heritage, pathways of development and general architecture that the constraints themselves become more interesting and more important in delimiting pathways of change than the selective force that may mediate change when it occurs. (*Spandrels*, 147)

They focused on the triangular areas at the tops of columns in St Mark's Church in Venice (Fig. 7.2). These areas are covered with exceptional mosaics; but the point of their argument is that the spandrels were not made – selected – for this purpose. The spaces are rather the side effects of the architecture needed to keep the roof in place. The spandrels seem to be designed for what they do. "Yet evolutionary biologists, in their tendency to focus exclusively on immediate adaptation to local conditions, do tend to ignore architectural constraints and perform just such an inversion of explanation." This opens the way for a good jab at – strong, hopefully fatal thrust at – an argument by E. O. Wilson, that Aztec human sacrifice is an adaptation to provide fresh meat.

> We strongly suspect that Aztec cannibalism was an 'adaptation' much like evangelists and rivers in spandrels, or ornamented bosses in ceiling spaces: a secondary epiphenomenon representing a fruitful use of available parts, not a cause of the entire system. To put it crudely: a system developed for other reasons generated an increasing number of fresh bodies; use might as well be made of them. Why invert the whole system in such a curious fashion and view an entire culture as the epiphenomenon of an unusual way to beef up the meat supply. Spandrels do not exist to house the evangelists. (584)

From here Gould and Lewontin launched into a general critique of Darwinism. "We wish to question a deeply engrained habit of thinking among students of evolution. We call it the adaptationist programme, or the Panglossian paradigm" – after Dr Pangloss of Voltaire's *Candide*, who saw everything as being part of the "best of all possible worlds." As an alternative, they turned to a continental position focusing on *Baupläne*, the underlying archetypes structuring organisms.

**Figure 7.2** Spandrel in San Marco church in Venice.

[This position] acknowledges conventional selection for superficial modifications of the Bauplan. It also denies that the adaptationist programme (atomization plus optimizing selection on parts) can do much to explain Baupläne and the transitions between them. But it does not therefore resort to a fundamentally unknown process. It holds instead that the basic body plans of organisms are so integrated and so replete with constraints upon adaptation . . . that conventional styles of selective arguments can explain little of interest about them. It does not deny that change, when it occurs, may be mediated by natural selection, but it holds that constraints restrict possible paths and modes of change so strongly that the constraints themselves become much the most interesting aspect of evolution. (594)

## Organicist Philosophy

Paralleling the science, we find that in philosophy there are those eager to promote a directed drive from within the organism as the main force for change. In 2009, the year when biologists were celebrating the 200th anniversary of Charles Darwin's birth, English philosopher John Dupré set the scene by referring in a sneering way to the undue "Darwinolatry." No lover of natural selection, he suggested, in the spirit of Sewall Wright, that genetic drift can do the necessary creative work. "Conceptually, selection and drift are quite different processes, but in practice they can be extremely difficult to separate. Once we see that the trajectory of a population through time is one in which adaptedness is always maintained, the conditions for existence are continuously met, it is very difficult to distinguish among the causes of this maintenance". He continues: "Where does adaptive change come from? A trivial but sometimes obfuscated point is that it never comes from natural selection," elaborating: "Selection cannot occur unless some other process provides alternatives to select from. It follows that any thesis about the power of natural selection to generate change implicitly presupposes a thesis about a process or processes that generate selectable change". By the time selection gets to the scene, the heavy lifting has been done, presumably by a non-random mutation. In some sense, the organism is driven by internal factors towards a desired end – acorn to oak, caterpillar to butterfly, monad to man.

Openly, Dupré unambiguously rejected the mechanical world picture in his 2012 writings. "There are powerful reasons for thinking that emancipation from the mechanistic paradigm is a precondition for true insight into the nature of biological processes" (*Processes*, 83). This opened the way to a turn to the organic world picture: "there are limits as to how far conventional mechanistic explanations can take us in understanding the dynamic stability of processes at this hierarchy of different levels. Such understanding will require models that incorporate both the capacities required by mechanistic or quasi-mechanistic constituents, and the constraints and causal influences provided by properties of the wider systems of which these constituents are parts" (203). Little surprise that reductionist thinking was rejected. "Traditional reductionist views of science, with their focus on 'bottom-up' mechanisms, do not suffice in the quest to understand top-down and circular causality and a world of nested processes".

American philosopher Thomas Nagel argued in much the same vein. The title of his book – *Mind and Cosmos: Why the Materialist Neo-Darwinian Conception of Nature Is Almost Certainly False* – tells all. Unlike Gould and Lewontin, Nagel agreed entirely with Darwin in finding the design-like nature of the organic world overwhelming. What he could not accept was what he saw as the random, undirected elements in the Darwinian picture. Like Dupré, Nagel was deeply suspicious of the machine metaphor perspective. He suggested that possibly "there are natural teleological laws governing the development of organization over time, in addition to laws of the familiar kind governing the behavior of the elements." He agrees that "This is a throwback to the Aristotelian conception of nature, banished from the scene at the birth of modern science. But I have been persuaded that the idea of teleological laws is coherent, and quite different from the intentions of a purposive being who produces the means to his ends by choice. In spite of the exclusion of teleology from contemporary science, it certainly shouldn't be ruled out a priori" (22).

To complete the story, turn to the respected American philosopher of mind, the late Jerry Fodor. He was so anti-Darwinian he hoped for a new paradigm; not, apparently, without reason to hope. "In fact, an appreciable number of perfectly reasonable biologists are coming to think that the theory of natural selection can no longer be taken for granted. This is, so far, mostly straws in the wind; but it's not out of the question that a scientific revolution – no less than a major revision of evolutionary theory – is in the offing ...". Like Nagel,

Fodor obsessed about the directionless nature of the Darwinian picture. "The present worry is that the explication of natural selection by appeal to selective breeding is seriously misleading, and that it thoroughly misled Darwin. Because breeders have minds, there's a fact of the matter about what traits they breed for; if you want to know, just ask them. Natural selection, by contrast, is mindless; it acts without malice aforethought." And so organicism starts to enter the picture, with its focus on the individual (rather than the group) as the cause of change. Embryological development. As Fodor and Piattelli-Palmarini explained:

> The slogan is the evolution of ontogenies. In other words, the whole process of development, from the fertilized egg to the adult, modifies the phenotypic effects of genotypic changes, and thus 'filters' the genotypic options that ecological variables ever have a chance to select from. (27)

Fodor explains:

> External environments are structured in all sorts of ways, but so, too, are the insides of the creatures that inhabit them. So, in principle at least, there's an alternative to Darwin's idea that phenotypes 'carry implicit information about' the environments in which they evolve: namely, that they carry implicit information about the endogenous structure of the creatures whose phenotypes they are.

And as organisms are self-programmed towards their ends, could this not also be the case for evolving species? Using an argument that, anticipating, upset Wallace – effective selection of any kind implies mind – Fodor continued:

> Darwin was too much an environmentalist. He seems to have been seduced by an analogy to selective breeding, with natural selection operating in place of the breeder. But this analogy is patently flawed; selective breeding is performed only by creatures with minds, and natural selection doesn't have one of those. The alternative possibility to Darwin's is that the direction of phenotypic change is very largely determined by endogenous variables. The current literature suggests that alterations in the timing of genetically controlled developmental processes is often the endogenous variable of choice; hence the 'devo' in 'evo-devo'.

Fifteen years earlier, a trio of non-Darwinian biologists, Gilbert, Opitz, and Raff, had made the same point.

> The homologies of process within morphogenetic fields provide some of the best evidence for evolution – just as skeletal and organ homologies did earlier. Thus, the evidence for evolution is better than ever. The role of natural selection in evolution, however, is seen to play less an important role. It is merely a filter for unsuccessful morphologies generated by development. Population genetics is destined to change if it is not to become as irrelevant to evolution as Newtonian mechanics is to contemporary physics. (Resynthesizing, 368)

## Against the Scientists

Before we end the argument and go out for a beer, let us see what response can be made by neo-Darwinians. Start with Sewall Wright and the shifting balance theory – adaptive peaks surrounded by non-adaptive valleys, genetic drift enabling populations to go down into the valleys and then up to a higher peak, some form of group selection passing the successful genes through the whole population. Since the theory was proposed, study after study, in the spirit of Dobzhansky, has shown that the theory is either irrelevant or not working. No one denies that there might be random factors in evolution, the founder principle for example. The question we want answered is whether drift is the causal process moving groups from one adaptive peak to another. There the evidence is less, to the point of non-existence. Starting with Dobzhansky, we have already seen how he discovered that fruit fly chromosome differences were not a function of drift, as he thought previously, but of selective forces to do with heat and humidity and so forth. There are many similar instances. Evolutionary biologist Jerry Coyne and colleagues, pointing out that Wright's prime example of genetic drift at work is of the flower-color polymorphism of the desert plant, *L. parryae*, have noted that there are two varieties, blue and white, the colors controlled by a single gene (that is, two alleles, one for blue and the other for white). Unfortunately for Wright, more recent and more detailed studies suggest that ratios of blue to white in populations remain stable, suggesting selection not drift. Moreover, "transplant experiments, studies of clines [groups with gradations from beginning to end] over short distances, and direct measurement of seed production all implied fairly strong selection between morphs, probably due to differences in soil moisture". Different temperatures could also be selective factors.

As has been said, no one is denying the possibility of genetic drift or that it might sometimes occur. Indeed, at the molecular level, where selection is ineffective, it is thought that drift can be very significant. This belief is used to date past events, like the break of human ancestors from the apes (Kimura 1968). This, however, is not our concern. We are focusing on the phenotypic, the physical level, where selection can be effective. The question is whether things all add up enough to say that drift is vital for moving groups of organisms from one form to another, better adapted. And there, selection is so often found to be a crucial factor, it is reasonable to conclude that, overall, the evidence for the full shifting balance theory (SBT) is slim to non-existent. Coyne and colleagues concluded:

> We do not doubt that the SB process might sometimes operate in nature. Given the multifarious nature of evolution, almost every conceivable scenario must occasionally occur. Many species are subdivided, genetic drift must sometimes oppose natural selection, and some evolutionary transitions probably involved peak shifts. Nevertheless, we have found no compelling evidence that Wright's SBT accounts for the evolution of a single adaptation, much less a significant proportion of adaptations, in nature.

Next up, C. H. Waddington with his "genetic assimilation," supposedly in some sense leading to directed mutations. Remember, he subjected fruit flies to extreme, shocking conditions, which caused anomalies (lack of cross-veins) in growth (see Fig. 7.1). He then selected from the anomalies. With a very few generations, he got the anomalies without need of environmental disturbance, the shocks; all in the cause he imbibed from Whitehead's organicist view of the world. Today's organicist biologists venerate Waddington. As Kevin Laland and colleagues put it: "It has long been argued that phenotypic accommodation could promote genetic accommodation if environmentally induced phenotypes are subsequently stabilized and fine-tuned across generations by selection of standing genetic variation, previously cryptic genetic variation or newly arising mutations". All importantly, this process means that evolution can be rapid as organisms take advantage of new opportunities.

As expected, neo-Darwinians think little of this. Surely what is happening is that the cross-vein is controlled by several alleles. Normally, taken individually, they all function correctly, producing the cross-veins. When heat-shocks are applied, this changes things sufficiently that some no longer function and those with these kinds of genes have no cross-vein. Selection on those that do not show a cross-vein produces descendant organisms with several such genes. These, taken as a whole, have sufficient power to produce wings without cross-veins, even in the absence of heat-shocks. One might add that in real life – as Waddington acknowledged – that there is no reason to think that cross-veinedness is of adaptive advantage, so little reason to think there would be selection for it.

This is not to deny that plasticity could in theory be important in evolution. Selection-supporting biologist Douglas Futuyma wrote:

> Therefore, phenotypic plasticity could be said to truly play a leading role (with genes as followers) if an advantageous phenotype were to be triggered by an environment that really is novel for the species lineage, an environment that its recent ancestors did not experience and which, therefore, had not exerted natural selection. Of course, it is possible that a novel environment – a new pesticide, for example – could evoke a developmental effect that happens to improve fitness, just as it is possible that a random DNA mutation improves fitness.

He added, more pessimistically: "But no theory leads us to expect such an effect to be especially likely." Moreover, he pointed out that often – usually? – new environments lead to maladaptation. "A guppy (*Poecilia reticulata*) population exhibited evidently maladaptive plastic changes in expression of many genes when reared in a novel environment (one that lacked predatory fish); descendants of this population adapted to a predator-free environment by precisely opposite evolutionary changes in gene regulation."

## Spandrels

And so to Gould and Lewontin. On the basis of Darwin's ongoing use of Lamarckism, they argued that they were the true Darwinians – ecumenical about causes – not the selection-above-everything crew. Under this umbrella, they tout archetypes, *Baupläne*. These mean constraints, and, as Darwin said, and as Gould and Lewontin quote, the "constraints restrict possible paths and modes of change so strongly that the constraints themselves become much the most interesting aspect of evolution." However, what Darwin also said, and what Gould and Lewontin do not quote, is: "It is generally acknowledged that all organic beings have been formed on two great laws—Unity of Type, and the Conditions of Existence", concluding "the law of the Conditions of Existence is the higher law; as it includes, through the inheritance of former variations and adaptations, that of Unity of Type" (*Origin*, 206). Darwin would not be alone in his skepticism. Looking at today's actual science, it simply is not true that constraints, or whatever, render impossible a selection-based analysis of phenomena. To take but one example among very many, David Reznick and his associates have undertaken an ongoing (four-decade) study of Trinidadian guppies. The questions they asked are totally within the Darwinian paradigm, seeing whether selection – particularly in response to different predators – could and did affect the morphology and growth and more of guppies in different areas. To quote from the abstract of one of scores of studies, published in 2001:

> Prior research has demonstrated a strong association between the species of predators that co-occur with guppies and the evolution of guppy life histories. The evolution of these differences in life histories has been attributed to the higher mortality rates experienced by guppies in high-predation environments. Here, we evaluate whether there might be indirect effects of predation on the evolution of life-history patterns and whether there are environmental differences that are correlated with predation. (*Life-history*, 12)

They continue:

> We found that high-predation environments tend to be larger streams with higher light levels and higher primary productivity, which should enhance food availability for guppies. We also found that guppy populations from high-predation environments have many more small individuals and fewer large individuals than those from low-predation environments, which is caused by their higher birth rates and death rates. Because of these differences in size distribution, guppies from

high predation environments have only one-fourth of the biomass per unit area, which should also enhance food availability for guppies in these localities. Guppies from high-predation sites allocate more resources to reproduction, grow faster, and attain larger asymptotic sizes, all of which are consistent with higher levels of resource availability.

What did they conclude?

[G]uppies from high-predation environments experience higher levels of resource availability in part because of correlated differences in the environment (light levels, primary productivity) and in part as an indirect consequence of predation (death rates and biomass density). These differences in resource availability can, in turn, augment the effect of predator-induced mortality as factors that shape the evolution of guppy life-history pattern.

All rather impressive, and surely strong support for a selection-based view of the world of organisms. Yes, indeed, but the trouble here is that Gould and Lewontin are, to be deservedly uncharitable, somewhat slippery in their arguments. They would agree entirely about the merits of the guppy study but respond that that was not quite relevant to what they were claiming. They were not talking about the everyday workings of selection but of rather special events – with the implication that these are very much more important than everyday workings. They were talking about the fundamental breakthroughs of evolution. Remember, although they acknowledged conventional selection for superficial modifications of the *Bauplan*, their position denied that selection can explain *Baupläne* and the transitions between them (*Spandrels*, 594). In other words, selection may be adequate for guppy birthrates, but when it comes to things like the vertebrate archetype, it is out of its depth. Here we must turn to constraints. In the words of Gould and Lewontin: "the basic body plans of organisms are so integrated and so replete with constraints upon adaptation ... that conventional styles of selective arguments can explain little of interest about them."

This is obviously a challenge we must speak to. This we will do, when we get to the chapter-ending discussion of evo-devo; but note already how much is being left to selection. It's not just Trinidadian guppy lifestyles. As Owen pointed out in 1849, the vertebrate archetype covers humans, horses, dolphins, birds, and bats – not to mention the dinosaurs and many other now extinct species. It was selection that made humans bipedal and intelligent, and horses fast and vegetarian, and dolphins happy denizens of the deep, and. . . In short, despite the somewhat condescending tone of the spandrels paper, natural selection is a hugely important part of the evolutionary process – arguably the most important part. For the moment, let us leave this valuation. We shall return to it.

## Against the Philosophers

Start with the question of progress. Organicism is inherently progressivist: acorn to oak. Dupré is explicit: "evolutionary continuity with the rest of life doesn't mean that there may not be features of human existence quite radically different from any found outside the human sphere" (Dupré 2003, 75–76). This is in total opposition to the Darwinian approach, where victory in the struggle for existence is a relative matter. Who wins, wins. To quote the immortal words of paleontologist John J. Sepkoski Jr., "I see intelligence as just one of a variety of adaptations among tetrapods for survival. Running fast in a herd while being as dumb as shit, I think, is a very good adaptation for survival." That we ourselves judge humans as superior is one thing. Anyone who says otherwise has been taking animal rights activist Peter Singer too seriously. It is quite another thing to claim that humankind, judged strictly in biological terms, is superior to every possible virus that might strike it.

Darwin recognized this. In an early notebook he wrote: "in my theory there is no absolute tendency to progression" (*Charles Darwin's Notebooks*, N 47). At other times, however, his Victorian personality led him to more optimistic conclusions. Great revolutionary but no rebel.

> Thus, from the war of nature, from famine and death, the most exalted object which we are capable of conceiving, namely, the production of the higher animals, directly follows. There is grandeur in this view of life, with its several powers, having been originally breathed into a few forms or into one; and that, whilst this planet has gone cycling on according to

the fixed law of gravity, from so simple a beginning endless forms most beautiful and most wonderful have been, and are being, evolved. (490)

Darwin did not just leave things there. In the third edition of the *Origin*, published in 1861, he appealed to what we today would call "arms races" – lines of organisms competing against each other and developing ever-more powerful adaptations, ending with human intelligence.

If we look at the differentiation and specialisation of the several organs of each being when adult (and this will include the advancement of the brain for intellectual purposes) as the best standard of highness of organisation, natural selection clearly leads towards highness; for all physiologists admit that the specialisation of organs, inasmuch as they perform in this state their functions better, is an advantage to each being; and hence the accumulation of variations tending towards specialisation is within the scope of natural selection. (134)

Richard Dawkins is a modern-day enthusiast for this kind of thinking. He points out that competing nations use ever-more sophisticated computer technology. In the animal world, Dawkins sees the evolution of bigger and bigger brains. We won!

A more recent progress-producing process starts with the notion of a niche, an ecological space occupied by organisms. Fish occupy the ocean niche; birds, the air niche. British paleontologist Simon Conway Morris argued in 2003 that not every possible niche is a functioning niche – presumably the bowel of a volcano is one such non-functioning niche. The need for usable niches constrains the route of evolution. Dumbo remains a Walt Disney fantasy. However, since we humans exist, there is clearly an intelligence niche. Plausibly, had we not found it, sooner or later, some organism would have found it, and, while not necessarily identical to us – they could have green skin and six digits – some kind of humanoid would have appeared, close enough to make it reasonable to talk of progress. "We may be unique, but paradoxically those properties that define our uniqueness can still be inherent in the evolutionary process. In other words, if we humans had not evolved then something more-or-less identical would have emerged sooner or later" (*Life's Solution*, 196).

These things come in threes. A third kind of putative progress-producing mechanism also has Darwinian echoes. From an early Darwin notebook (early 1838, just after he came up with the idea of selection): "The enormous number of animals in the world depends on their varied structure & complexity" (*Charles Darwin's Notebooks*, E 95). Linking this to selection, Darwin goes on: "it is quite clear that a large part of the complexity of structure is adaptation" (E 97). In this spirit, paleontologist Daniel McShea and his colleague at Duke University, philosopher Robert Brandon, argued that an updated version of this insight does that which is necessary. They posited what, in 2010, they called the "zero-force evolutionary law" (ZFEL for short), explaining: "In any evolutionary system in which there is variation and heredity, in the absence of natural selection, other forces, and constraints acting on diversity or complexity, diversity and complexity will increase on average" (3). It would seem things just naturally keep complexifying, and given enough time, humans, or human-like beings, will appear.

None of these natural-selection-fueled, progress-producing suggestions is really that convincing. It is generally agreed that evolutionary arms races do occur, but it is far from obvious that they are necessarily going to end with humankind. The best-known example is that of shellfish getting ever-tougher shells to protect against boring predators, which in turn get ever more efficient as the shells get thicker and tougher. Not much about intelligence here. The Sepkoski objection is fatal. Intelligence is costly. You need large amounts of high-energy protein, namely the bodies of other animals. There were no vegans in the Pleistocene. So, intelligence is not predestined to appear. Often, a better way of succeeding was to go the route of KISS – keep it simple, stupid. As evolutionist George Williams pointed out in 1966, sometimes simple is better than complex. As a general design, jet engines are a lot simpler than internal combustion engines. Humans appeared. They did not necessarily appear.

Intelligence niches seem no less successful. Apart from anything else, many would question the claim that niches seem to exist objectively, waiting to be invaded. Lewontin argued in 2002 that organisms create their niches as much as find them. He wrote: "organisms not only determine what elements of the outside world are relevant to them by peculiarities of their shapes and metabolism, but they actually construct, in the literal sense of the word, a world around themselves" (*Triple Helix*, 54). Who is to say that some animal is going to create a human welcoming niche? It could be a long wait. There could be niches far above us. Is it not presumptuous to assume we are the highest or only kind of intelligence? To quote J. B. S. Haldane from 1927: "Now, my own suspicion is that the universe is not only queerer than we suppose, but queerer than we can suppose."

The same goes for complexity. Agree that, on the whole, things are going to get more complex. Sooner or later, these things happen. But, as we have just seen, for all that earlier we saw that Sepkoski might be thinking in these terms, paradoxically it is he who points out that complexity does not at once spell better adapted. There seems no inevitability to the evolution of brains. And in any case, the quote from McShea and Brandon explicitly allows that, unlike Darwin, they are not appealing to natural selection. In fact, it is not entirely certain that they themselves believe what they have said.

> We conclude what is known about the history of life offers little evidence for the ZFEL for complexity. A longterm increase in the mean has not been demonstrated, but if in fact it occurred, it would be consistent with a number of possible underlying mechanisms. The ZFEL predicts a strong drive, but no such drive has been shown, and indeed the stable minimum suggested by impressionistic assessments argues for the opposite, a weak drive or none at all. (*First Law*, 84)

Note again that I am not denying that we humans think we are superior to warthogs. Or that philosophers think they are the brightest people on campus. The person who says otherwise is either lying or very peculiar. It is just that this is a judgment that we make. It is not something given to us by science. The Darwinian is happy with this. The organicist is, or should be, uncomfortable.

Let us turn to the other philosophers. Nagel is entitled to argue for an Aristotelian perspective. He still must tell us how he thinks a teleological world system can function. He is an atheist, so at once he rules out the Platonic Designer, the Form of the Good or the Demiurge. He cannot really embrace undiluted Aristotelianism. If there is no Unmoved Mover to which all strives, one still needs to know why things strive and why they strive as they do, towards ends. At the risk of seeming irate, one would have more respect for thinkers like Nagel if they showed any evidence of having read – glanced at – a piece of modern, professional, evolutionary theorizing – Resnick on guppies, for instance. Nagel is on a par with the New Atheists when they criticize Islam. As the saying goes: ignorance is (rather noisy, self-important) bliss.

## Is Evo-Devo Relevant?

We come to the argument from development. Here supposedly, thanks to insights from the newly articulated branch, or subbranch, of evolutionary developmental biology – dubbed "evo-devo" – comes the decisive blow of organicism against mechanistic Darwinism, with its central course of natural selection. This is the claim of the philosopher Jerry Fodor. Even more, it is the central theme of the Spandrels paper. Natural selection cannot explain basic body plans – archetypes – because they are so integrated and so replete with constraints upon adaptation, they are beyond the scope of Darwin's central cause.

As we turn to this topic, a couple of background points. First, Darwin himself was fully aware of the importance of development. Although he had not the conceptual tools to deal with the issue, he warned in the *Origin* that "correlation of growth will have had a most important influence in modifying various structures" (196). He added: "Correlation of growth has no doubt played a most important part, and a useful modification of one part will often have entailed on other parts diversified changes of no direct use" (199). Second, when the neo-Darwinians were putting their theory together, they were aware that they were treating the developing organism like a black box, or more accurately a sausage machine. Genes (genotypes) go in at one end, and organisms (phenotypes) come out at the other. It was not that development was thought unimportant. To think that selection enthusiasts were about to downplay or deny – or feel fear at – evolutionary development, could not be farther from the truth. First things first.

Some of the discoveries about development have been truly staggering, particularly "deep homology" at the molecular level. In his inimitable style, Ernst Mayr, in talking (in the 1960s) about the importance of homology for working out relationships, loftily assured us that no one in their right mind would think of looking for homologies between fruit flies and humans. He lived long enough to eat his words. The similarity of the molecules (and their order) of fruit flies and humans, when it comes to the production and order of basic body parts, is mind blowing. Frogs and mice too! What is going on? Tempting though it might be, at times, to think of one's graduate students as lower forms of animal, truly we humans are not fruit flies, or frogs or mice for that matter. What is happening is obvious. There are shared genes – "homeotic" genes – that control and order and vary development. "*Hox*" genes determine how the body parts are put together sequentially in an organism. How these genes play out in the fruit fly is shown in Fig 7.3a. The incredible and rather humbling extent to which humans follow a similar pattern controlled by similar genes is shown in Fig. 7.3b. Of course, overall, what ends up in the fruitfly world is

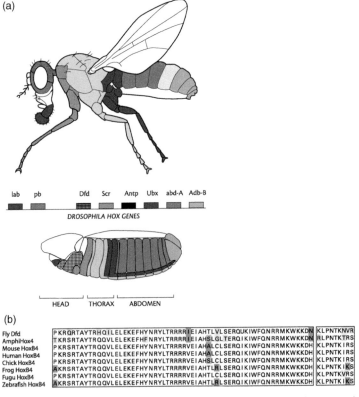

**Figure 7.3**    a, Drosophila *hox* genes. b, *Hox* genes of humans and Drosophila compared.

going to be very different from what ends up in the human world. It is the same story as occurs in, let us say, the human and the bird. They share the same basic archetype, and then evolutionary causes prepared us for a hunter-gatherer existence and the bird for life in the air.

As an aside, a thought which should not be dismissed is that something like this is overwhelming evidence of design – or design-like processes. In the absence of an intelligent being, which neither Fodor nor Gould seem inclined to invoke, it would seem that natural selection is already being flagged as an incredibly important part of the process. None of this is to deny that non-selective factors are going to be important. Some of the factors will be external. "Niches" are the ecological quarters in which organisms reside – for instance, the niches of squirrels are trees – and "niche construction" is where organisms themselves help to create their quarters, for instance, beavers building dams. The niche within which an organism finds itself, or helps create, is going to put restraints on possible development. If there are no animals for prey, not much point in developing adaptations for meat-eating. If the environment is very cold, little place for above-ground, naked mole-rats. Perhaps factors like these are going to bring in intelligent beings that direct evolution (in an organic fashion) from within rather than, as with selection, from without. As Kevin Laland and colleagues describe it: "Niche construction should be recognized as an evolutionary process because it imposes a statistical bias on the direction and mode of selection that ensue, and hence on the speed and direction of evolution. By systematically creating and reinforcing specific environmental states, niche construction directs evolution along particular trajectories" (Niche construction, 2). Just so. All this is true, but hardly news to the Darwinian and even less worrying. Where did the intelligence come from in the first place, if not from selection? Organisms good at finding or creating niches survived and reproduced; those that were not, did not.

What about more internal factors? Again, no one will deny that there are such factors and that these do put constraints on the power of selection. As Gould pointed out, no mammal has developed wheels for transport, even though they would obviously be of great adaptive value. The laws of anatomy and physiology are absolute. Likewise, there are going to be no elephant-like creatures with legs as nimble as those of the antelope. The laws of physics tend to be terrible spoilsports. Moreover, the very processes of development will themselves be open to constraints – things they cannot do, however much selection may try. One well-regarded experimental study – done on butterflies with two spots on their wings – confirmed this. Two significant features of these spots are size and color. Artificial selection showed that the size of spots could, under selection, vary on the same wing. However, artificial selection failed entirely to separate the colors of the spots. They stayed coordinated (Fig. 7.4). Seemingly, the genes would allow one kind of change, but not the other. As evolutionary biologist Wallace Arthur explains:

> So, the general conclusion seems to be that in some cases of developmental bias, the bias will affect the direction of evolutionary change, while in others it will not, because selection can override it. In other words, sometimes the direction of evolutionary change is determined by selection alone, while sometimes it is the result of the interplay between natural selection and the structure of the available variation. (85)

There are indeed constraints. It has always been known that organisms are in many respects jerry-built. George C. Williams – a fanatical

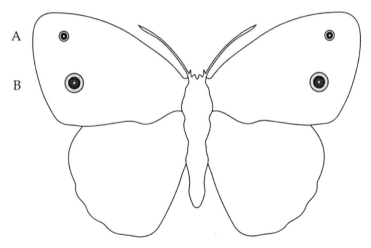

**Figure 7.4** Butterfly spots.

Darwinian – pointed this out. The male urogenital system could only have been designed by Rube Goldberg. The point is that constraints are challenges, not roadblocks. It is not a question of being absolutely fast – just faster than the chap next to you.

## Archetypes

Take up the Gould/Lewontin critique. Does this necessarily mean that, outside regular (selection-driven) evolution, there are going to be mega events, like the evolution of the vertebrate archetype? Once the archetype is on the table – less metaphorically, structuring a group of organisms in a hitherto unknown way – there are huge possibilities. True, this needs to be within the constraints of the archetype – there are no six-legged mammals – but there is still a great deal you can do with four legs. Gould and Lewontin asked: How do you get the archetype on the table in the first place? Don't these mega-evolutionary events call for non-selection-guided saltations? Getting so much, so quickly?

Not necessarily. First note that mega events are judged after the fact. That the vertebrate archetype proved so successful was hardly a given from the start. It might have been wiped out by a random event like an ice age. Or it might have been that gravity was so strong – Jupiter? – that it simply could not function. If you say that the whole point of the archetype was that it could function given our gravity, then you start to make it even better designed. And that makes even more pressing the source of this design.

An Intelligent Designer? If you rule this out, then the only viable option is selection working on regular variations. The fact that we judge something to be a mega event is a judgment after the fact. It does not speak to the causal process bringing it about. No one is denying that sometimes changes, perhaps in genes controlling development, can bring about major changes. But these are changes occurring on the already-existing, not changes to the absolutely new. At least, not successful changes to the absolutely new. As so very often, J. B. S. Haldane had it right. He wrote in 1932:

> To sum up, it would seem that natural selection is the main cause of evolutionary change in species as a whole. But the actual steps by which

individuals come to differ from their parents are due to causes other than selection, and in consequence evolution can only follow certain paths. These paths are determined by factors which we can only very dimly conjecture. Only a thoroughgoing study of variation will lighten our darkness. (*Causes*, 142–3)

Evo-devo is an incredibly exciting extension of Darwinian selection theory, offering challenges and opportunities of mouth-watering promise. It is far from a threat.

# 8 Natural Selection and Its Discontents

When a new cause is introduced into science, as often as not it is accepted without trouble. Few, if any, had worries about the Watson–Crick double helix and the subsequent working out of the genetic code. Genetics was put on a molecular causal basis. However, it is not uncommon for there to be opposition. Huygens' wave theory of light was an outsider for nearly two centuries. Sometimes worries are ongoing. One doubts that, as long as there are those interested in mental health, Freud's Oedipus Complex is going to be happily accepted by all. There have been, continue to be, and probably always will be disputes, often bitter, about its causal status. As we have seen, natural selection did not have an altogether easy birth. But as time went by, things seem to have improved. Newton and Leibniz all over again:

> I was interested the other day in reading the Life of Newton by Brewster to find that Leibnitz actually attacked the Law of Gravity as 'subversive of all Natural Religion'!! He further attacked Newton for having used gravity 'an occult quality' to explain the motions of the Planets.— Newton answered that it is philosophy to explain movement of wheels of clock, though the cause of descent of the weight could not be explained. This seems to me rather to bear on what you say of Nat. selection not being proved as a vera causa.— (Darwin, Letter to Asa Gray, February 24 1860)

It surely seems true that, today, natural selection is accepted easily and readily by working evolutionists. To take but one example, the already-mentioned, deservedly highly praised work on the evolution of Darwin's finches in the Galapagos, by the husband-and-wife team, fellows of the Royal Society, Peter and Rosemary Grant. In one celebrated study, there were changes in the food supply of one species of finch due to a severe El Niño event. Large seeds became scarce, and so there was a knock-on effect of natural selection bringing on small beak sizes. The moral? As the Grants explained in 1993:

> The evolutionary change documented in the population of *G. forlis* on Daphne serves as a model for what has presumably happened countless times on a larger timescale in the past: small evolutionary changes in quantitative traits caused by natural selection under changing environmental conditions. (*Finches*, 116)

Here is a second example, from a 1995 paper about microevolutionary changes in the finches: "The best scope for a predictive theory of evolution lies in the area of genetics, because the mathematical machinery has been developed for the precise prediction of evolutionary change caused by selection" (*Predicting*, 241). And a third example, from a 2003 abstract about species diversity: "Darwin's finches on the Galápagos Islands are particularly suitable for asking evolutionary questions about adaptation and the multiplication of species: how these processes happen and how to interpret them." And how does it all happen? "Key factors in their evolutionary diversification are environmental change, natural selection, and cultural evolution. A long-term study of finch populations on the island of Daphne Major has revealed that evolution occurs by natural selection when the finches' food supply changes during droughts" (965). If this is not a causal explanation, I do not know what would be.

And yet. If there is a hated – and feared – phenomenon in science, it is natural selection. The charges go all the way from it is masquerading as a real cause to that (not entirely consistently) it led to Hitler. Concluding this short book, I shall look at some of these charges, starting with challenges to its status as genuine scientific knowledge (epistemology) and then on to its supposed moral iniquities (ethics). Although there is no one linking theme, it is notable how often discussions revolve around metaphor. This is perhaps not surprising because, although metaphor is a vital aspect of science, it has always caused uneasiness. For all that the Scientific Revolution depended on the embrace of the machine root metaphor, at once there were philosophers and others telling us how dangerous metaphors are and how they should be avoided and eliminated. Few can beat Samuel Parker, a member of the Royal Society, who in 1661 spoke of metaphors, as Andrew Reynolds describes, as "wanton and luxuriant phantasies climbing up into the Bed of Reason, [that] do not only defile it by unchast and illegitimate Embraces, but instead of real conceptions and notices of things impregnate the mind with nothing but Ayerie and Subventaneous Phantasmes" (2).

## Survival of the Fittest

Today, as we have seen, there is a more balanced attitude to metaphor – indispensable, often useful, but always dangerous. To get things going, let us start with a golden oldie: the "survival of the fittest." If natural selection is anything, it is a metaphor. Humans literally select. Nature does not literally select. But in Darwinian theory, metaphorically, nature selects. And many were uncomfortable with this, starting with the co-discoverer of natural selection, Alfred Russel Wallace. In fact, although Wallace certainly got hold of the idea of natural selection, from the first he was not very keen on the analogy/metaphor. In his original essay, "On the tendency of varieties to depart indefinitely from the original type," he acknowledged that many assumed there is an analogy between domestic and wild organisms: "it is the object of the present paper to show that this assumption is altogether false" (Wallace 1858). Domestic forms, if changed, have a tendency to return to the original type. This is not true of wild organisms. Although Wallace went on to accept Darwin's take on things, including the term natural selection, he was never completely easy. Is Darwin not back with a quasi-Platonic view of a mastermind, with intelligence – or rather, Intelligence – doing the real work? On July 2, 1866, Wallace wrote at length to Darwin, about how many people seemed to think that natural selection brought in a thinking designer:

> The two last cases of this misunderstanding are, 1st. The article on "Darwin & his teachings" in the last "Quarterly Journal of Science", which, though very well written & on the whole appreciative, yet concludes with a charge of something like blindness, in your not seeing that "Natural Selection" requires the constant watching of an intelligent "chooser" like man's selection to which you so often compare it;—and 2nd., in Janet's recent work on the "Materialism of the present day", reviewed in last Saturday's "Reader", by an extract from which I see

that he considers your weak point to be, that you do not see that "thought & direction are essential to the action of 'Nat. Selection'." The same objection has been made a score of times by your chief opponents, & I have heard it as often stated myself in conversation.

Now I think this arises almost entirely from your choice of the term "Nat. Selection" & so constantly comparing it in its effects, to Man's selection, and also to your so frequently personifying Nature as "selecting" as "preferring" as "seeking only the good of the species" &c. &c. To the few, this is as clear as daylight, & beautifully suggestive, but to many it is evidently a stumbling block. I wish therefore to suggest to you the possibility of entirely avoiding this source of misconception in your great work, (if not now too late) & also in any future editions of the "Origin", and I think it may be done without difficulty & very effectually by adopting Spencer's term (which he generally uses in preference to Nat. Selection) viz. "Survival of the fittest."

Wallace continued in the same vein.

This term is the plain expression of the facts,—Nat. selection is a metaphorical expression of it—and to a certain degree indirect & incorrect, since, even personifying Nature, she does not so much select special variations, as exterminate the most unfavourable ones. (Letter from Wallace to Darwin, July 2, 1866)

Darwin was sufficiently appreciative of this letter that, for the later editions of the *Origin*, he added Spencer's term as an alternative. The title of the chapter dealing with natural selection, in the 1869 fifth edition, was altered to NATURAL SELECTION, OR THE SURVIVAL OF THE FITTEST.

Two questions. Was Darwin right to do this? Did Darwin have to do this? Answering the first, he was possibly – probably – right. Clearly, people were misinterpreting him, and it was best to introduce Spencer's term to avoid this. Answering the second is not so straightforward. For all the problems with natural selection, survival of the fittest is not exactly guilt-free. It is reproduction that matters, not survival, and it is fitter that counts, not fittest. Chased by a bear, it is not a matter of beating everyone; just the chap next to you, also running for his life. Today, somewhat more robustly than Darwin and

Wallace, very secular evolutionists have no qualms about speaking of "natural selection" (and less inclination to use "survival of the fittest"). Richard Dawkins, of all people, uses the term "natural selection" fifty-six times in *The Selfish Gene.* As in: "Once upon a time, natural selection consisted of the differential survival of replicators floating free in the primeval soup. Now, natural selection favours replicators that are good at building survival machines, genes that are skilled in the art of controlling embryonic development." I leave as an exercise for the reader the deconstruction of "floating free" and "primeval soup" and "survival machines." More pertinent to us: no, there are no references, in *The Selfish Gene,* to the "survival of the fittest."

Unfortunately, survival of the fittest did not vanish without leaving a bad philosophical taste in the mouth. It gave rise to countless charges that natural selection is no real cause. It is simply a tautology. Survival of the fittest. But who are the fittest? Those that survive! Hence, natural selection cashes out as: those that survive are those that survive. True, but not of great explanatory value. Karl Popper made this point:

> Take "Adaptation". At first sight natural selection appears to explain it, and in a way it does, but it is hardly a scientific way. To say that a species now living is adapted to its environment is, in fact, almost tautological. Indeed we use the terms "adaptation" and "selection" in such a way that we can say that, if the species were not adapted, it would have been eliminated by natural selection. Similarly, if a species has been eliminated it must have been ill adapted to the conditions. Adaptation or fitness is *defined* by modern evolutionists as survival value, and can be measured by actual success in survival: there is hardly any possibility of testing a theory as feeble as this.

To be fair, Popper was not entirely negative. He did allow that Darwinism is a "metaphysical research programme." But still! That would hardly require the forty years of work the Grants have put in on the Galapagos. Obviously, there is something wrong. If Popper were right, then genetic drift would be a contradiction in terms, for it postulates that the fittest do not always survive, and even the harshest critics of genetic drift do not want to claim that this is the case. Too easy a victory. The fault, in reasoning like Popper, is that what survives or reproduces in one case might not be so successful in another case.

Generally, having wings is a good thing for insects. Those that have wings survive, and those that don't, don't. But when it comes to oceanic islands, then having wings is not necessarily a good thing. You can too easily get blown out to sea. So, selection gets rid of your wings, and you have to forage in different ways. What survives, survives. True. But what it takes to survive is another matter – an empirical matter. There is no tautology here. Natural selection is more than a truism.

## Natural Selection as Cause

But even if you save natural selection here, have you given it the place that the evolutionists seem to think it has, as the cause of evolutionary change? We have seen the debate over Darwin's *vera causa* claim. Now we have an updated version: Is natural selection a cause? A band of thinkers, philosophers and others, think not. "Natural selection is not a cause of evolution" philosophers Mohan Matthen and André Ariew have suggested (201). Australian psychologist Ben Bradley writes:

> Nowadays, we often see natural selection framed as a cause, process, or mechanism. *Origin*'s rich metaphorical language did indeed sometimes cast natural selection as an all-powerful agency, incessantly working to improve the world – as Darwin soon came to regret. But *the argument* of the book framed things differently, making natural selection a law resulting from what it called the struggle for life, not a causal mechanism or process. After *Origin* came out, Darwin repeatedly rebutted critics who argued that his book had failed because it had not shown natural selection to be a *true cause* ('vera causa') of species change. *Origin* called natural selection a *general law*, 'leading to the advancement of all organic beings – namely, multiply, vary, let the strongest live and the weakest die.' By *law*, Darwin meant merely a 'sequence of events as ascertained by us.' In this, natural selection was like Newton's law of gravity, he said. The law of gravity gave new coherence to various observable sequences of events, such as falling apples, tidal flows, and orbiting planets. But the law of gravity did *not* specify what mechanisms caused these events – a conundrum which taxes physicists to this day. Likewise, certain processes *caused* the sequences of events brought together under the law of

natural selection. In this sense, natural selection did not refer to a causal power. It described effects. (*Psychology*, 76)

Some philosophers, such as Charles Pence, think that this kind of criticism still applies today. We supposedly have the statisticalist view (correct) versus the causalist view (incorrect): "the statisticalists argue that natural selection and genetic drift (other factors in evolution are rarely considered) are merely convenient summaries of the genuinely causal events taking place in the lives of individual organisms. When organisms eat, fight, mate, and die, these causal events are what powers evolution. We, as theorists, then make the decision to abstract away from these fine-grained details and build evolutionary explanations using terms like selection, fitness, and drift" (1.2). In other words, the causal issues occur down on the ground, as it were. If, say, one of the Grants' populations of finches gets (as they said) a small beak because the food is now small grains not large ones, it is not so much selection which is bringing it on, but that some individuals with larger beaks are not eating big grains and dying, and others with smaller beaks eating small grains and living. The cause is the beak crushing grains – or not. Natural selection is simply keeping score, as does the Dow Jones average. The Dow Jones does not make things (cause things to) happen. It is just statistics about what did happen.

This might well be true of species selection. One is just keeping score of what has happened. With respect to individual selection, one is tempted to say, in the face of a highly successful empirical science, that only philosophers would think this way. This is Richard Dawkins' position. "I mean it as a compliment when I say you could almost define a philosopher as someone who won't take common sense for an answer" (83). Apparently, however, psychologists can get infected too. The problem starts with metaphor. Natural selection is a metaphor based on the thoughts and actions of breeders. The question is what gets transferred over in the mechanism, and what does not/need not. As the poet Robert Frost said in 1931 about the machine metaphor – no one thinks the world had to have "a pedal for the foot, a lever for the hand, or a button for the finger." Does the natural selection metaphor demand something akin to the breeders and their thoughts and actions? Vitalists thought it did. Listen to Henri Bergson, on his vital force, the *élan vital*.

> [Take] the idea we started from, that of an original impetus of life, passing from one generation of germs to the following generation of germs through the developed organisms which bridge the interval between the generations. This impetus, sustained right along the lines of evolution among which it gets divided, is the fundamental cause of variations, at least of those that are regularly passed on, that accumulate and create new species. (*Creative Evolution*, 87–8)

To be honest, this sounds more like a Lamarckian situation than a Darwinian selective one – which is perhaps no great surprise given that Bergson stands in the Romantic organicist tradition and was much influenced by Spencer – but the key point is that something living is driving things, something akin to the living breeder. Darwinism won't have this, and so the critics are led to conclude that this means you cannot have a cause. But is this not a bit too quick? Turn to the thinking of the greatest philosopher to write in the English language. David Hume pointed in new philosophical directions that the Darwinian follows. Most importantly, causes are not to be thought of things "out there," but rather as things provided by our minds to conceptualize – to make sense of – experience. Hume wrote: "It is easy to observe, that in tracing this relation, the inference we draw from cause to effect, is not derived merely from a survey of these particular objects, and from such a penetration into their essences as may discover the dependance of the one upon the other. There is no object, which implies the existence of any other if we consider these objects in themselves, and never look beyond the ideas which we form of them." He continued in his *Treatise of Human Nature*:

> It is therefore by EXPERIENCE only, that we can infer the existence of one object from that of another. The nature of experience is this. We remember to have had frequent instances of the existence of one species of objects; and also remember, that the individuals of another species of objects have always attended them, and have existed in a regular order of contiguity and succession with regard to them. Thus we remember, to have seen that species of object we call flame, and to have felt that species of sensation we call heat. We likewise call to mind their constant conjunction in all past instances. Without any farther ceremony, we call the one cause and the other effect, and infer the existence of the one from that of the other.

That is precisely the Darwinian's point. There is no "thing," natural selection, "out there." We see the laws in motion. We see the effect of these laws. We speak of "cause." The differential reproduction and survival of organisms leads to change. We conceptualize this differential reproduction as the "cause." By using the metaphor of selection, we can get to work explaining. Seeing a beak crushing small nuts is simply seeing a beak crushing small nuts. With the metaphor, we can understand what is going on. Selection is pushing the population of finches towards small-beakiness, because (in the absence of big nuts) only those that can crush small nuts will survive and reproduce. Kant was absolutely right: the concept of a thing in itself has a practical use "for guiding research into objects of this kind and thinking over their highest ground in accordance with a remote analogy with our own causality in accordance with ends." Thinking this way, pretending that there is intention out there, is heuristic. Where Kant was wrong was in thinking this cannot be part of real science and for that reason biology is second rate. To the contrary, the metaphor of natural selection is a part of good science. That is all there is to it. And, with this, we can conclude with the Darwinian that there is a cause of evolution: natural selection.

## Progress

Already we have had much to say about a third epistemological-type objection, namely that natural selection is but one of a number of causes, and by no means obviously the most important one. This is the overriding theme of the Gould and Lewontin paper on spandrels. They wanted to counter Edward O. Wilson's Darwin-inspired forays into human evolution, and they did this by attacking him obliquely: natural selection is not all it is made out to be. "We feel that the potential rewards of abandoning exclusive focus on the adaptationist programme are very great indeed," continuing: "A pluralistic view could put organisms, with all their recalcitrant, yet intelligible, complexity, back into evolutionary theory" (*Spandrels*, 597). As we have seen, the Gould/ Lewontin pluralistic view makes much of *Baupläne*, archetypes. And openly they agreed that the view of evolution they are endorsing falls within the Germanic neo-Platonic (although they certainly don't use that term) way of thinking. Organicist. And, as a consequence, holistic. In the "unfairly maligned" approach they endorsed, "evolutionists have never been much attracted to the Anglo-American penchant for atomizing organisms into parts and trying to explain each as a direct adaptation" (593).

What is not emphasized is the extent to which the organicist position is inherently progressive, leading up to humans (Fig. 8.1). As it happens, a few years earlier, in a major organicist-friendly work linking individual development with group development – ontogeny with phylogeny – Gould had written:

> I have been trying to deemphasize the traditional arguments of morphology while asserting the importance of life-history strategies. In particular, I have linked accelerated development to r-selected regimes and I have identified retarded development as a common trait of K strategists... I have also tried to link K selection to what we generally regard as "progressive" in evolution, while suggesting that r selection generally serves as a brake upon such evolutionary change. I regard human evolution as strong confirmation of these views. (*Ontogeny*, 504)

Given the inherently non-progressivist nature of Darwinian evolution, one might well conclude that the spandrels attack was not a one-off against natural selection, but a manifestation of a general organicist belittlement of selection. Ironically, it was just at this time that Gould was turning strongly against biological progress. In large part, this was because such progress was built into the hated system of Edward O. Wilson – for all that Wilson has been painted the archetypical Darwinian, there were always organicist elements in his thinking. He has ever been an enthusiast about Herbert Spencer. Gould's feeling (not entirely without justification) was that Wilson's ideal human was a grey-suited, New York businessman in the 1950s.

Moving on from Wilson directly, as or more important in Gould's turn from progress was the conviction that notions of human progress always leave some groups behind, and – expressing a sentiment for which we have already seen good evidence – it was this kind of philosophy that led to the Third Reich and the Final Solution. It gave a pseudo-scientific veneer to anti-Semitism. So, in typically flamboyant style, Gould – proudly Jewish – in 1988 excoriated biological progress as "a noxious, culturally embedded, untestable, nonoperational, intractable idea that must be replaced if we wish to understand the

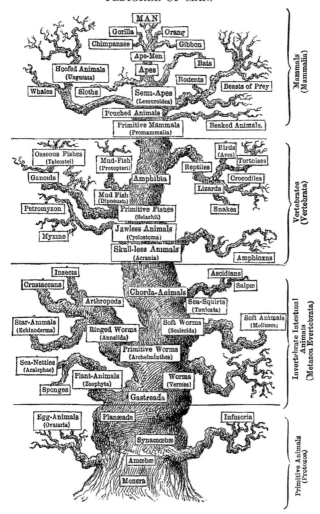

**Figure 8.1**   The tree of life as drawn by Ernst Haeckel in *The Evolution of Man* – whereas Darwin's tree in the *Origin* is schematic, showing branching, this tree is intended as a true picture of life's history.

patterns of history" (319). That humans evolved is contingent, not necessary. Had the asteroid not hit Earth and wiped out the dinosaurs about 66 million years ago, there is little chance that there would be humans today. "Since dinosaurs were not moving toward markedly larger brains, and since such a prospect may lie outside the capabilities of reptilian design ... we must assume that consciousness would not have evolved on our planet if a cosmic catastrophe had not claimed the dinosaurs as victims. In an entirely literal sense, we owe our existence, as large and reasoning mammals, to our lucky stars" (318).

## Social Darwinism

F. Scott Fitzgerald wrote that "The test of a first-rate intelligence is the ability to hold two opposing ideas in mind at the same time". In a not-entirely atypical fashion, Gould was running with the hare and hunting with the hounds – somewhat unsuccessfully. If you want to run down natural selection morally, don't hang progress on it; if you want deny progress, then don't think you are going after natural selection. This is not at all to dispute that there are many who would fault natural selection morally. The usual move is to focus on the struggle for existence and to argue that selection leads to a brutal, ethically bereft society. Thomas Henry Huxley was good on this: "Man, the animal, in fact, has worked his way to the headship of the sentient world, and has become the superb animal which he is, in virtue of his success in the struggle for existence" (*Evolution and Ethics*, 51), adding: "For his successful progress, throughout the savage state, man has been largely indebted to those qualities which he shares with the ape and the tiger; his exceptional physical organization; his cunning, his sociability, his curiosity, and his imitativeness; his ruthless and ferocious destructiveness when his anger is roused by opposition." Nevertheless, "in proportion as men have passed from anarchy to social organization, and in proportion as civilization has grown in worth, these deeply ingrained service-able qualities have become defects. After the manner of successful persons, civilized man would gladly kick down the ladder by which he has climbed. He would be only too pleased to see 'the ape and tiger die'" (52).

What is truly dreadful is that, unlike Huxley, many thought that this was all a good idea. The philosophy of "Social Darwinism" found enthusiastic converts. Remember Herbert Spencer on the foolishness of those who would combat the fact that "under the natural order of things, society is constantly excreting its unhealthy, imbecile, slow, vacillating, faithless members." Andrew Carnegie, the Scottish-born American industrialist, sang the same song. "The law of competition may be sometimes hard for the individual, [but] it is best for the race, because it insures the survival of the fittest in every department" (*Gospel*, 655). This was nothing on General Friedrich von Bernhardi, sometime member of the German General Staff. "War is a biological necessity," and hence: "Those forms survive which are able to procure themselves the most favourable conditions of life, and to assert themselves in the universal economy of nature. The weaker succumb." Without war, progress is at an end: "inferior or decaying races would easily choke the growth of healthy budding elements, and a universal decadence would follow." Von Bernhardi could give Thrasymachus in the *Republic* a run for his money: "Might gives the right to occupy or to conquer. Might is at once the supreme right, and the dispute as to what is right is decided by the arbitrament of war. War gives a biologically just decision, since its decision rests on the very nature of things" (quoted by Crook, *Darwinism*, 83).

We are on the way to Hitler. As Reichard Weikart, Intelligent Design enthusiast who has made a career out of tarring Darwin with the Hitler brush, argues: "In his writings and speeches Hitler regularly invoked Darwinian concepts, such as evolution (*Entwicklung*), higher evolution (*Höherentwicklung*), struggle for existence (*Existenzkampf* or *Daseinskampf*), struggle for life (*Lebenskampf*), and selection (*Auslese*)" (*Darwinism*, 541). And there is evidence of this. From Hitler's *Mein Kampf*:

All great cultures of the past perished only because the originally creative race died out from blood poisoning.

The ultimate cause of such a decline was their forgetting that all culture depends on men and not conversely; hence that to preserve a certain culture the man who creates it must be preserved. This preservation is bound up with the rigid law of necessity and the right to victory of the best and stronger in this world.

Those who want to live, let them fight, and those who do not want to fight in this world of eternal struggle do not deserve to live. (1, chapter 11)

Perhaps we would have been better off if we had agreed that natural selection is not a cause, and so could not be responsible for any of this! Unfortunately, in real life there are no easy ways out.

Start with the fact that Spencer was not a Darwinian and never put natural selection in the center of the picture. We saw that, when he was talking about "excreting" its unhealthy members, he was not talking about selection but about spurs to action so that people would strive to move upwards through Lamarckian processes. Carnegie likewise was less interested in the non-survival of the unfit and more in the survival of the fit. That is why he founded public libraries so the poor-but-bright child could go and find literature to better themselves for free. I come from the same socio-economic background as Herbert Spencer – Margaret Thatcher too. I spent many happy hours of my childhood in the Carnegie library in Walsall, in the British Midlands. I leave it for the reader to decide whether Carnegie invested his money wisely.

Von Bernhardi is a rather different case. He really did believe in the virtues of selection against the weak. "That struggle eliminates the weak and used-up nations, and allows strong nations possessed of a sturdy civilisation to maintain themselves and to obtain a position of predomin-ant power until they too have fulfilled their civilising task and have to go down before young and rising nations" (*Germany*, 26). How much of his thinking is due to Darwin is a moot point. He loathed and detested the British. There were plenty of German sources on which he could and did draw. Thus Hegel: "I have remarked elsewhere, 'the ethical health of peoples is preserved in their indifference to the stabilisation of finite institutions; just as the blowing of the winds preserves the sea from the foulness which would be the result of a prolonged calm, so also corruption in nations would be the product of prolonged, let alone "perpetual" peace'" (*Elements*, 324; the reference to perpetual peace was a jab at Kant, who had written about its possibility).

Hitler too had his own non-Darwinian sources. As historian Bob Richards has shown, highly influential was the English-born Houston Stewart Chamberlain. A fanatical Wagnerian – he married the composer's daughter – his *Foundations of the Nineteenth Century* published in 1899 portrayed recent history as a battle between Aryans – "great, heavenly radiant eyes, golden hair, the body of a giant, harmonious musculature" and so forth – and the Jew – "materialistic, legalistic, limited in imagination, intolerant, fanatical, and with a tendency toward utopian economic schemes" (214–15). And this points to the fact that Hitler was not so much writing about the struggle between races as the need to eliminate Jews. Science was always going to be a support rather than a foundation.

Notice how we are directed to a more general point about the Nazi use of evolutionary biology. Germany always has had quality science. In the nineteenth and early twentieth century, it was the world leader, until America started to flex its intellectual muscles. Of course, you are going to find people knowledgeable about natural selection and able and willing to use it in contexts that today we find morally deplorable. Weikart lists a number of scientists in the Third Reich who made use of selection for ideological ends. In 1941, Heinz Brücher, in an official Nazi Party publication, *Nationalsozialistische Monatshefte*, laid it all out:

> The hereditary health of the German Volk and of the Nordic-Germanic race that unites it must under all circumstances remain intact. Through an appropriate compliance with the laws of nature, through selection and planned racial care it can even be increased. The racial superiority achieved thereby secures for our Volk in the harsh struggle for existence an advantage, which will make us unconquerable. (Brücher 1941, quoted by Weikart, *Darwinism*, 51)

Even more depressing is the Wannsee Protocol – recorded by Adolf Eichmann, the minutes of the Wannsee Conference, held early in 1942, when the "Final Solution" was discussed and agreed upon. An understanding of natural selection is presupposed, as Jeffery O'Connell and I have explained:

> Under proper guidance, in the course of the final solution the Jews are to be allocated for appropriate labor in the East. Able-bodied Jews, separated according to sex, will be taken in large work columns to these areas for work on roads, in the course of which action doubtless a large portion will be eliminated by natural causes.

> The possible final remnant will, since it will undoubtedly consist of the most resistant portion, have to be treated accordingly, because it is the product of natural selection and would, if released, act as the seed of a new Jewish revival. (See the experience of history.) (49–50)

One can say, with good reason, that Darwin is hardly responsible for any of this. The artillery in the First World War, blasting each other across no-man's land, were relying on Galileo's laws of motion; but we would not hold Galileo responsible for the horrors of the Battle of the Somme. Likewise, we are hardly going to hold Darwin responsible for the Wannsee Protocol, written over a hundred years since he discovered natural selection.

## Values in Science

This said, given the uses to which it could be and was put, one might still say that natural selection is inherently evil. Red flags should go up at once. This objection supposes that natural selection – a scientific concept – has absolute value in itself. But, as we have seen, under the machine metaphor, within science, there are no such absolute values. Here it distinguishes itself from the organic metaphor that does see such values – monad to man. So, what is this objection all about? It is obvious. You don't use natural selection just on its own. There are always lots of other factors to bring in. Take this passage from an article by the Grants in 2003: "What Darwin's finches can teach us about the evolutionary origin and regulation of biodiversity". It makes the point precisely. Natural selection is in play but against a background of environmental change.

> In this article we survey the evidence from field studies of the ecological causes of diversification. The explanation for diversification involves natural selection, genetic drift, introgressive hybridization, and genetic as well as cultural evolution. Linking all these factors are the frequent and strong fluctuations in climatic conditions, between droughts on the one hand and extremely wet (El Niño) conditions on the other. An important conclusion of this study is that environmental change is an observable driving force in the origin of new species. (965)

It is the same in the Nazi case. Go back to Heinz Brücher and Weikart: "In a 1935 article Heinz Brücher praised Ernst Haeckel for paving the way for the Nazi regime. In addition to mentioning Haeckel's advocacy of eugenics and euthanasia, Brücher highlighted Haeckel's role in promoting human evolution. [See Fig. 8.1.] Brücher reminded his readers that Haeckel's view of human evolution led him to reject human equality and socialism", continuing: "In Brücher's view human evolution is an essential ingredient of racial ideology, not a hindrance to it". "Ingredient." That tells it all. Eggs are an essential ingredient for waffles; but, without flour and butter, you get an omelet, not waffles. (Without maple syrup, you are better off with cornflakes.)

Nevertheless, you might object that although something may in itself be value-free, the use to which we will put it shows value bias. A guillotine in itself is just a guillotine; but no one is ever going to use it for topping carrots. Likewise with natural selection, you might say. In itself, natural selection is value-free, but it lends itself to Nazi-supporting arguments. The analogy fails. No doubt there are other uses to which you might put a guillotine – as an exhibit in a torture chamber exhibit – but they are not obvious and, as like or not, as in this case, a derivation of the main function, chopping off human heads. Natural selection can be used in a huge number of cases, with moral implications the very opposite of National Socialism. There is, for example, massive evidence that skin color is the direct result of strong selective pressures – ten to twenty thousand years is enough time for a population to change to an optimal color, as anthropologist Nina Jablonski has explained. Humans in hot areas tended to lose bodily hair so they could sweat more easily and reduce their temperatures. But this opened them to direct sunlight, bringing on more exposure to ultraviolet (UV) light, which on the one hand is needed to help vitamin D produce the calcium needed for strong bones, but on the other hand strips away folic acid, needed for the production of healthy babies. Melanin, which produces dark skin, is a vital adaptation to protect the body from too much UV light. Farther north, with less sun, the need is to keep up the UV exposure so that vitamin D can do its job. Hence, people in the north tend to have lighter skin. Interestingly, putting a nice confirmatory cap on this explanation, people who eat lots of seafood can get their vitamin D directly from their diets. Hence indigenous people in Alaska and the north of Canada can and do have darker skins. These in turn protect them from too much UV light, due to reflection in the summer from snow and ice.

Notwithstanding cultural prejudices today, about the relative standings of blacks and whites, there simply is nothing in this explanation prejudicial to any group of humans. The outcome of natural selection is that it keeps humans alive and reproducing. Skin color variation is part and parcel of this process. There is nothing about differences in intelligence between groups. Indeed, one suspects that much of the motivation behind the studies about skin color was precisely to combat prejudice. Moreover, these skin-color studies are but one small part of what we are now learning, thanks to ancient DNA techniques and studies, about human history in the past twenty or so thousand years. There has been movement back and forth across continents as humans sought better living conditions and responded to natural events like Ice Ages. For instance, Britain was invaded by people who made pots named "beakers," about 4,000 years ago – with huge consequences. "The genetic impact of the spread of peoples from the continent into the British Isles in this period was permanent. British and Irish skeletons from the Bronze Age that followed the Beaker period had at most about 10 percent ancestry from the first farmers of these islands, with the other 90 percent from people like those closely associated with the Bell Beaker culture in the Netherlands" as geneticist David Reich has argued, (115). The people who built Stonehenge are not the people walking around it today. In short, all the guff about the "racial superiority" of the Volk, thanks to "the harsh struggle for existence," really starts to sound like outdated science.

## Darwin on Morality

If we grant all of this, still there remains a worry that natural science has less-than-moral consequences. Remember Huxley: "After the manner of successful persons, civilized man would gladly kick down the ladder by which he has climbed. He would be only too pleased to see 'the ape and tiger die.'" Huxley rarely got Darwin right, and there is no reason to think he did so here. It is true that the Darwinian view of life respects vigor and energy – the very things Huxley had in abundance. But this is a good thing. Remember *New Grub Street*. The brilliant novelist turns out to be quite inadequate as a human being and ends up dead. The less-talented man recognizes his own limitations but uses them for strength, not despair. He doesn't waste his time writing unpublishable novels. He sets to, gets the editorship of a distinguished journal – like *The Times Literary Supplement* – and, using his talents, is able to make a real contribution to society.

More generally, as Darwin made very clear in the *Descent of Man,* natural selection did not make us into lions or tigers. It made us into cooperators. As we have seen, since leaving the jungle long ago, our ancestors were hunter-gatherers, bands of about fifty people, on what has rather nicely been described as a five-million-year camping trip. Our strength, as Darwin saw, is in cooperation. We are not very fast. We are not very strong. We are hopeless about climbing trees. And so much more. But we were and are very clever primates. We discovered that, by working together, we could succeed with style, as it were. We developed adaptations to help this process – speech for instance. We are not special or divine in being peaceable – we are well adapted. Like being bipedal, this is no magic. Just natural selection.

Morality? Darwin saw how this came to be. It is a very good tool – adaptation – for enforcing sociality. You are tired and hungry and don't much want to share. But you know that you should. You don't always do the moral thing, but you do it often enough that it works.

> It must not be forgotten that although a high standard of morality gives but a slight or no advantage to each individual man and his children over the other men of the same tribe, yet that an advancement in the standard of morality and an increase in the number of well-endowed men will certainly give an immense advantage to one tribe over another. There can be no doubt that a tribe including many members who, from possessing in a high degree the spirit of patriotism, fidelity, obedience, courage, and sympathy, were always ready to give aid to each other and to sacrifice themselves for the common good, would be victorious over most other tribes; and this would be natural selection. (*Descent of Man*, vol. 1, 166)

You might think, as Elliott Sober thinks that Darwin is endorsing group selection. Not so! For a start, morality is not just to the advantage of others, but to your advantage. You scratch my back, and I will scratch yours:

Darwin wrote that "as the reasoning powers and foresight of the members [of a tribe] became improved, each man would soon learn from experience that if he aided his fellow-men, he would commonly receive aid in return" (*Descent of Man*, 1, 163). Today, biologists refer to this as "reciprocal altruism." This doesn't mean that, in an important sense, reciprocal altruism is not genuine altruism. If my genes trick me into thinking that I ought to help others, I am probably going to function a lot more efficiently than if I am constantly scheming and lying, pretending to be a friend, but truly always calculating. A society of people like this is not going to last long.

One should add that Darwin was inclined to think that tribes are groups of interrelated individuals, or at least individuals who think themselves inter-related. He endorsed an article by Herbert Spencer on tribes, where it is clearly argued that tribes think themselves united by a common ancestor, whether this be strictly true or not. "If 'the Wolf,' proving famous in fight, becomes a terror to neighbouring tribes, and a dominant man in his own, his sons, proud of their parentage, will not let fall the fact that they descended from 'the Wolf'; nor will this fact be forgotten by the rest of the tribe who hold 'the Wolf' in awe, and see reason to dread his sons." Indeed, the rest of the tribe will want to get onboard. "In proportion to the power and celebrity of 'the Wolf' will this pride and this fear conspire to maintain among his grandchildren and great-grandchildren, as well as among those over whom they dominate, the remembrance of the fact that their ancestor was 'the Wolf'" (*On ancestor worship*, 535 ). Darwin agreed: "names or nicknames given from some animal or other object to the early progenitors or founders of a tribe, are supposed after a long interval to represent the real progenitor of the tribe" (*Descent of Man*, 66, ft. 53). Morality for Darwin comes from something we have seen already being posited as the behind-the-scene force for bringing about the caste system in social insects: a kind of family selection – what, as we have seen, is today known as "kin selection" – driven in our case by our belief in our being part of a related community. In the words of today's evolutionists such as Gary Johnson: "we are evolutionarily primed to define 'kin' as those with whom we are familiar due to living and rearing arrangements. So genetically unrelated individuals can come to be understood as kin – and subsequently treated as such – if introduced into our network of frequent and intimate associations (for example, family) in an appropriate way" (133).

Let us leave things now. It need hardly be said how Darwin's account of the selection-fueled evolution of morality meshes perfectly with what we saw in Chapter 6 is today's thinking about hunter-gatherers. We humans took a very distinctive route. We gave up being lions and tigers, at least within our groups. We became social beings. And the evolution of morality was, in its way, as important as starting to speak. Say no more!

# Envoi

In 1866, Thomas Hardy, raised a sincere member of the Church of England, wrote his sonnet "Hap." It expressed the anxiety about – "fear of" is not too strong a term – the world into which natural selection has pitched us. No longer can we rely on a Good God to care for us, to suffer for us, to make possible eternal life. In the non-progressive world of Darwinian evolution, all is meaningless.

> If but some vengeful god would call to me
> From up the sky, and laugh: "Thou suffering thing,
> Know that thy sorrow is my ecstasy,
> That thy love's loss is my hate's profiting!"
>
> Then would I bear it, clench myself, and die,
> Steeled by the sense of ire unmerited;
> Half-eased in that a Powerfuller than I
> Had willed and meted me the tears I shed.
>
> But not so. How arrives it joy lies slain,
> And why unblooms the best hope ever sown?
> —Crass Casualty obstructs the sun and rain,
> And dicing Time for gladness casts a moan...
> These purblind Doomsters had as readily strown
> Blisses about my pilgrimage as pain.

This poem speaks to much that we have seen and discussed in this short book. Even the most secular among us feel that natural selection is not just another scientific concept, like the waves of light or even – dare one say it? – the force of gravitational attraction. It strikes right to the heart of our being. We humans are, together with all other living organisms, an end-product of a history without planned direction. Does this mean that, as Shakespeare said, life is "a tale told by an idiot, full of sound and fury, Signifying nothing"? That, as Camus said, life is "absurd"? Answering this question is left as an exercise for the reader. I can only reiterate what I said in my preface that, after more than fifty years as a scholar and teacher, I am still overwhelmed by what our reason and senses tell us about how our world works, and I can think of no greater meaning to life than, as I have done here, sharing that knowledge with others.

# Summary of Common Misunderstandings

Here are some of the oft-encountered misunderstandings of natural selection, issues that will be discussed in the text. My advice is that you read through these first and then, after reading the main text, read them again and see if you agree with what I say. I am a philosopher so I shall take it as a matter of success, not that you agree with everything I say, but that you now feel that you have sufficient grasp of the issues – or are motivated to turn to other sources to learn more of the issues – to come to conclusions that are yours, not mine. I shall think you are not taking me seriously if you agree with everything I say!

**Natural selection requires a conscious agent that selects** Natural selection is a metaphor, based on the selection practiced by breeders, for better farm animals, for fancier birds, for stronger fighting dogs. Human selection obviously implies a conscious agent; the metaphor of natural selection directs us to think as if something similar is happening in nature. This does not imply that nature is conscious or directed by a conscious being. There may be a God or Intelligent Designer behind it all, but that claim is not part of the science.

**Survival of the fittest is a better metaphor than natural selection** "Survival of the fittest" is the term used by Herbert Spencer for natural selection. It was urged on Charles Darwin by the co-discoverer of natural selection, Alfred Russel Wallace, on the grounds that it was less likely to lead to anthropomorphism. Darwin agreed and introduced it into later editions of the *Origin*. Unfortunately, it leads to the charge that natural selection is a tautology – those that survive are those that survive – and it is rarely used by today's professional biologists.

**Evolution by natural selection implies that evolution is progressive** Today's evolutionists, using natural selection, are mechanists, in the sense that they accept the root metaphor implying that the world is machine-like, operating endlessly under unchanging laws. This precludes value notions like progress, the belief that things improve, from lesser to greater. Hence, there can be no claim, based on selection, that humans are the most important of creatures. A rival root metaphor that the world is organism-like sees evolution as a progressive unfurling, acorn to oak. This metaphor was embraced by the Romantics and is promoted by their followers, like Herbert Spencer and Henri Bergson and today's philosophers like Thomas Nagel. Humans are seen as the apotheosis of evolution.

**Darwin's and Wallace's theories of natural selection are essentially the same** Darwin and Wallace hit independently on the idea and significance of natural selection. However, they differed in three important ways. First, Wallace was never comfortable with the analogy between human selection and natural selection. He therefore urged the alternative term: "survival of the fittest." Second, whereas Darwin was an individual selectionist, Wallace was a group selectionist. He thought that hybrid sterility occurred for the good of the parent species, a claim Darwin denied strongly. Third, Wallace became a spiritualist, believing that non-natural forces were involved in human evolution. Appalled, in reaction, Darwin wrote the *Descent of Man,* arguing that natural processes can explain our origins.

**Natural selection was not accepted until the modern synthesis in the 1930s, leading to what has been described as the eclipse of Darwinism** After the *Origin*, and for the next fifty-plus years, natural selection had a mixed reception. Professional biologists, working mainly on morphology, or tracing phylogenies – the histories of organisms – had little need of selection, and so ignored it. This neglect was compounded by the lack of an adequate theory of heredity and the claim by physicists that the world is too young for a slow process like natural selection to have had a major effect. However, in the world of fast-breeding organisms, insects, selection was a big success. Bates's theory of mimicry attests to that. Also, in the world of culture, poetry and fiction, natural selection and its corollary, sexual selection, were instantly appreciated and used with gusto and perceptive understanding.

**Biological constraints make natural selection redundant** Notoriously, in their "Spandrels" paper, Stephen Jay Gould and Richard Lewontin claimed that organisms' underlying ground plans – archetypes or *Baupläne* – constrain the action of natural selection, which is reduced to a mere clean-up role. This is simply not true. When they first appeared, the *Baupläne* were hardly *Baupläne*. They were selection-produced forms, of value for what they could do now, without thought of their future value. Moving to the future, the organisms that evolved from the primitive forms, sharing the *Baupläne*, had many different adaptations. Birds, wings for flying; dolphins, fins for swimming; horses, legs for running; humans, hands and arms for grasping. How did these adaptations occur, if not through natural selection?

**Natural selection leads to Social Darwinism and then on to Hitler and National Socialism** This charge is based on the fact that natural selection starts with the struggle for existence and therefore promotes "nature red in tooth and claw." But, as Darwin makes clear in the *Origin*, "struggle" covers a lot of action, and sometimes you get more by cooperating than fighting. Some of the most notorious supposed Social Darwinians, like the American industrial baron Andrew Carnegie, were philanthropists, in his case sponsoring public libraries. Hitler drew more on home-grown German sources and, while there were undoubtedly National Socialist scientists who used natural selection to "prove" Aryan superiority, this hardly makes natural selection pernicious. There have been scientists who drew on natural selection to show that skin-color origins have nothing to do with intelligence and hence racial bias is based on bad science. As part of mechanistic science, in itself natural selection is value-free.

**Natural selection is anti-religion and a major reason why, today, many find life meaningless, without purpose** Natural selection did challenge older, religious views, for instance about Adam and Eve and original sin, but so also did other things like Higher Criticism. Many Christians and other religious believers found they could reconcile their science with their religion and that indeed the interaction enriched both sides. And those who now reject religion find there is still meaning to life, enriched rather than depleted, thanks to scientific ideas like natural selection. It is only with a greater understanding of human nature and its origins that we can start on a genuine (and successful) quest for meaning. We must have the will and integrity to stand on our own two feet and not rely on deities, real or apparent.

# References

## Chapter 1

Bowler, P. J. 1975. The changing meaning of 'Evolution'. *Journal of the History of Ideas* 36: 95–114.

Darwin, C. 1859. *On the Origin of Species by Means of Natural Selection, or the Preservation of Favoured Races in the Struggle for Life.* London: John Murray.

Evans, E. J. 2001. *The Forging of the Modern State: Early Industrial Britain, 1783–1870* 3rd Ed. Harlow, Essex: Longman.

Malthus, T. R. [1826] 1914. *An Essay on the Principle of Population* 6th Ed. London: Everyman.

Owen, R. 1843. *Lectures on the Comparative Anatomy and Physiology of the Invertebrate Animals.* London: Longman, Brown, Green and Longmans.

Owen, R. 1860. *Palaeontology or a Systematic Summary of Extinct Animals and their Geological Relations.* Edinburgh: Adam and Charles Black.

Ruse, M. 1979a. *The Darwinian Revolution: Science Red in Tooth and Claw.* Chicago: University of Chicago Press.

Smith, A. [1776] 1937. *The Wealth of Nations.* New York: Modern Library.

## Chapter 2

Agricola, G. [1556] 1950. *De Re Metallica.* (Translators H. C. Hoover and L. C. Hoover) New York: Dover.

Bacon, F. [1605] 1868. *The Advancement of Learning*. Oxford: Clarendon Press.

Barnes, J. (editor). 1984. *The Complete Works of Aristotle*. Princeton: Princeton University Press.

Boyle, R. [1686] 1996. *A Free Enquiry into the Vulgarly Received Notion of Nature*. (Editors E. B. Davis and M. Hunter ) Cambridge: Cambridge University Press.

Boyle, R. [1688] 1966. A Disquisition about the Final Causes of Natural Things. *The Works of Robert Boyle*. (Editor T. Birch) 5: 392–444. Hildesheim: Georg Olms.

Browne, J. 1995. *Charles Darwin: Voyaging. Volume 1 of a Biography*. London: Jonathan Cape.

Bury, J. B. [1920] 1924. *The Idea of Progress; An Inquiry into its Origin and Growth*. London: MacMillan.

Cooper, J. M. (editor) 1997. *Plato: Complete Works*. Indianapolis: Hackett.

Darwin, C. 1859. *On the Origin of Species by Means of Natural Selection, or the Preservation of Favoured Races in the Struggle for Life*. London: John Murray.

Darwin, C. 1987. *Charles Darwin's Notebooks, 1836–1844*. (Editors P. H. Barrett, P. J. Gautrey, S. Herbert, D. Kohn, and S. Smith), Ithaca, NY: Cornell University Press.

Darwin, E. [1794–1796] 1801. *Zoonomia; or, The Laws of Organic Life*. 3rd ed. London: J. Johnson.

Darwin, E. 1803. *The Temple of Nature*. London: J. Johnson.

Descartes, R. [1637] 1964. Discourse on Method. *Philosophical Essays*, 1–57. Indianapolis: Bobbs-Merrill.

Dijksterhuis, E. J. 1961. *The Mechanization of the World Picture*. Oxford: Oxford University Press.

Henderson, L. J. 1917. *The Order of Nature*. Cambridge, MA: Harvard University Press.

Herschel, J. F. W. 1830. *Preliminary Discourse on the Study of Natural Philosophy*. London: Longman, Rees, Orme, Brown, Green, and Longman.

Hume, D. [1779] 1963. Dialogues Concerning Natural Religion. *Hume on Religion*. (Editor R. Wollheim ) 93–204. London: Fontana.

Kant, I. [1790] 1928. *The Critique of Teleological Judgement*. (Translator J. C. Meredith) Oxford: Oxford University Press.

Kepler, J. [1619] 1977. *The Harmony of the World*. (Translators E. J. Aiton, A. M. Duncan, and J. V. Field) Philadelphia: American Philosophical Society.

Kuhn, T. 1962. *The Structure of Scientific Revolutions*. Chicago: University of Chicago Press.

Lakoff, G. and M. Johnson 1980. *Metaphors We Live By*. Chicago: University of Chicago Press.

Lovejoy, A. O. 1936. *The Great Chain of Being*. Cambridge, MA: Harvard University Press.

Lyell, C. 1830–1833. *Principles of Geology: Being an Attempt to Explain the Former Changes in the Earth's Surface by Reference to Causes now in Operation*. London: John Murray.

Lyell, C. 1881. *Life, Letters and Journals of Sir Charles Lyell, Bart*. London: John Murray.

Paley, W. [1802] 1819. *Natural Theology (Collected Works: IV)*. London: Rivington.

Pepper, S. C. 1942. *World Hypotheses: A Study in Evidence*. Berkeley: University of California Press.

Rudwick, M. J. S. 2005. *Bursting the Limits of Time*. Chicago: University of Chicago Press.

Ruse, M. 1975a. Charles Darwin and artificial selection. *Journal of the History of Ideas* 36: 339–50.

Ruse, M. 1996. *Monad to Man: The Concept of Progress in Evolutionary Biology*. Cambridge, MA: Harvard University Press.

Ruse, M. 2005a. *The Evolution-Creation Struggle*. Cambridge, Mass.: Harvard University Press.

Ruse, M. 2017. *On Purpose*. Princeton, NJ: Princeton University Press.

Ruse, M. 2021a. *A Philosopher Looks at Human Beings*. Cambridge: Cambridge University Press.

Ruse, M. 2021b. The Scientific Revolution. *The Cambridge History of Atheism*. (Editors S. Bullivant and M. Ruse ) Cambridge: Cambridge University Press.

Sebright, J. 1809. The art of improving the breeds of domestic animals. Letter addressed to the Right Hon. Sir Joseph Banks, K.B. London: Privately published.

Sedley, D. 2008. *Creationism and its Critics in Antiquity*. Berkeley: University of California Press.

Whewell, W. 1837. *The History of the Inductive Sciences (3 vols)*. London: Parker.

## Chapter 3

Anon. 1860a. Natural selection. *All the Year Round* 3: 293–99.

Anon. 1860b. Species. *All the Year Round* 3: 174–78.

Anon. 1861. Transmutation of species. *All the Year Round* 4: 519–21.

Bates, H. W. 1862. Contributions to an insect fauna of the Amazon Valley. *Transactions of the Linnean Society of London* 23: 495–515.

Bates, H. W. [1863] 1892. *The Naturalist on the River Amazon*. London: John Murray.

Bowler, P. J. 1976. *Fossils and Progress*. New York: Science History Publications.

Bowler, P. J. 1983. *The Eclipse of Darwinism: Anti-Darwinism Evolution Theories in the Decades around 1900*. Baltimore: Johns Hopkins University Press.

Bowler, P. J. 1988. *The Non-Darwinian Revolution: Reinterpreting a Historical Myth*. Baltimore, MD: Johns Hopkins University Press.

Bowler, P. J. 2013. *Darwin Deleted: Imagining a World Without Darwin*. Chicago: University of Chicago Press.

Burchfield, J. D. 1975. *Lord Kelvin and the Age of the Earth*. New York, NY: Science History Publications.

Burroughs, E. R. [1912] 1914. *Tarzan of the Apes*. Chicago: McClurg.

Comstock, J. H. 1893. Evolution and taxonomy. *The Wilder Quarter Century Book*, 37–114. Ithaca: Comstock Publishing.

Darwin, C. 1859. *On the Origin of Species by Means of Natural Selection, or the Preservation of Favoured Races in the Struggle for Life*. London: John Murray.

Darwin, C. 1868. *The Variation of Animals and Plants Under Domestication*. London: Murray.

Darwin, C. 1871. *The Descent of Man, and Selection in Relation to Sex*. London: John Murray.

Dixon, E. S. 1862. A vision of animal existences. *Cornhill Magazine* 5, no. 27: 311–18.

Duncan, D. (editor). 1908. *Life and Letters of Herbert Spencer*. London: Williams and Norgate.

Eliot, G. [1876] 1967. *Daniel Deronda*. London: Penguin.

Gissing, G. [1891] 1976. *New Grub Street*. London: Penguin.

Hull, D. L. 1979. The limits of cladism. *Systematic Zoology* 28: 416–40.

Hull, D. L. 1988. *Science as a Process*. Chicago: University of Chicago Press.

Huxley, T. H. 2009. *Evolution and Ethics with a New Introduction* (Editor M. Ruse ) Princeton: Princeton University Press.

James, H. 1873. Middlemarch. *Galaxy* 15: 424–8.

Jenkin, F. 1867. Review of 'The origin of species'. *The North British Review* 46: 277–318.

Kellogg, V. L. 1905. *Darwinism Today: A Discussion of the Present-Day Scientific Criticism of the Darwinian Selection Theories, together with a Brief Account of the Principle Other Proposed Auxiliary and Alternative Theories of Species-Forming*. New York: Henry Holt.

Lucas, J. R. 1979. Wilberforce and Huxley: A legendary encounter. *Historical Journal* 22: 313–30.

Müller, J. F. T. 1879. Ituna and Thyridia: A remarkable case of mimicry in butterflies. *Proceedings of the Entomological Society of London*: 20–29.

Naden, C. 1999. *Poetical Works of Constance Naden*. Kernville, CA: High Sierra Books.

Patterson, C. 1994. "Null or minimal models". In: *Models in Phylogeny Reconstruction*. (Editors R. W. Scotland and D. J. Siebert). Systematics Association Special Volume: 173–92.

Platnick, N. I. 1977. Review of concepts of species. *Systematic Zoology* 26: 97.

Platnick, N. I. 1979. Philosophy and the transformation of cladistics. *Systematic Zoology* 28: 538–46.

Poulton, E. B. 1890. *The Colours of Animals*. London: Kegan Paul, Trench, Truebner.

Richards, R. J. 2003. *The Romantic Conception of Life: Science and Philosophy in the Age of Goethe*. Chicago: University of Chicago Press.

Ruse, M. 1979a. *The Darwinian Revolution: Science Red in Tooth and Claw*. Chicago: University of Chicago Press.

Ruse, M. 1996. *Monad to Man: The Concept of Progress in Evolutionary Biology*. Cambridge, MA: Harvard University Press.

Spencer, H. 1851. *Social Statics: Or, the Conditions Essential to Human Happiness Specified, and the First of Them Developed*. London: Chapman.

Spencer, H. 1852. A theory of population, deduced from the general law of animal fertility. *Westminster Review* 1: 468–501.

Spencer, H. 1860. The social organism. *Westminster Review* LXXIII: 90–121.

Spencer, H. 1864. *Principles of Biology*. London: Williams and Norgate.

## Chapter 4

Allen, G. E. 1978. *Thomas Hunt Morgan: The Man and His Science*. Princeton, NJ: Princeton University Press.

Allison, A. C. 1954a. Protection by the sickle-cell trait against subtertian malarial infection. *British Medical Journal* 1: 290.

Allison, A. C. 1954b. The distribution of the sickle-cell trait in East Africa and elsewhere and its apparent relationship to the incidence of subtertian malaria. *Transactions of the Royal Society of Tropical Medical Hygiene* 48: 312.

Bateson, W. 1902. *Mendel's Principles of Heredity. A Defence, with a Translation of Mendel's Original Papers on Hybridisation*. Cambridge: Cambridge University Press.

Bowler, P. J. 1989. *The Mendelian Revolution: The Emergence of Hereditarian Concepts in Modern Science and Society*. London: The Athlone Press.

Dobzhansky, T. 1937. *Genetics and the Origin of Species*. New York: Columbia University Press.

Dobzhansky, T. 1943. Temporal changes in the composition of populations of *Drosophila pseudoobscura* in different environments. *Genetics* 28: 162–86.

Dobzhansky, T. 1951. *Genetics and the Origin of Species* 3rd Ed. New York: Columbia University Press.

Fisher, R. A. 1930. *The Genetical Theory of Natural Selection*. Oxford: Oxford University Press.

Ford, E. B. 1931. *Mendelism and Evolution*. London: Methuen.

Ford, E. B. 1964. *Ecological Genetics*. London: Methuen.

Haldane, J. B. S. 1932. *The Causes of Evolution*. New York: Cornell University Press.

Hardy, G. H. 1908. Mendelian proportions in a mixed population. *Science* 28: 49–50

Majerus, M. E. N. 1998. *Melanism: Evolution in Action*. Oxford: Oxford University Press.

Paley, W. [1802] 1819. *Natural Theology (Collected Works: IV)*. London: Rivington.

Provine, W. B. 1971. *The Origins of Theoretical Population Genetics*. Chicago: University of Chicago Press.

Provine, W. B. 1986. *Sewall Wright and Evolutionary Biology*. Chicago: University of Chicago Press.

Sheppard, P. M. 1958. *Natural Selection and Heredity*. London: Hutchinson.

Tutt, J. W. 1890. Melanism and melanochroism in British lepidoptera. *The Entomologist's Record, and Journal of Variation* 1, no. 3: 49–56.

Wright, S. 1931. Evolution in Mendelian populations. *Genetics* 16: 97–159.

Wright, S. 1932. The roles of mutation, inbreeding, crossbreeding and selection in evolution. *Proceedings of the Sixth International Congress of Genetics* 1: 356–66.

Wright, S. 1945. Tempo and mode in evolution: a critical review. *Ecology* 26: 415–19.

## Chapter 5

Bell, C. 1834. *The Hand: Its Mechanism and Vital Endowments as Evincing Design*. London: William Pickering.

Darwin, C. 1839. Observations on the parallel roads of Glen Roy, and of other parts of Lochaber in Scotland, with an attempt to prove that they are of marine origin. *Philosophical Transactions of the Royal Society of London*, 129: 39–81.

Darwin, C. 1842. *The structure and distribution of coral reefs. Being the first part of the geology of the voyage of the Beagle, under the command of Capt. Fitzroy, R. N. during the years 1832 to 1836*. London: Smith Elder.

Darwin, C. 1862. *On the Various Contrivances by which British and Foreign Orchids are Fertilized by Insects, and On the Good Effects of Intercrossing*. London: John Murray.

Darwin, C. 1909. *The Foundations of the Origin of Species: Two Essays Written in 1842 and 1844*. (Editor F. Darwin ) Cambridge: University of Cambridge Press.

Dawkins, R. 1976. *The Selfish Gene*. Oxford: Oxford University Press.

Delisle, R. G. (editor) 2021. *Natural Selection: Revisiting its Explanatory Role in Evolutionary Biology*. Cham, Switzerland: Springer Nature.

Gray, A. 1876. *Darwiniana*. New York, NY: D. Appleton.

Herschel, J. F. W. 1830. *Preliminary Discourse on the Study of Natural Philosophy*. London: Longman, Rees, Orme, Brown, Green, and Longman.

Herschel, J. F. W. 1841. Review of Whewell's *History and Philosophy*. *Quarterly Review* 135: 177–238.

Huxley, T. H. [1859] 1893. The Darwinian hypothesis. *The Times*, December 26.

Irwin, D. E., S. Bensch, and T. D. Price. 2001. Speciation in a ring. *Nature* 409: 333–7.

Lyell, C. 1830–1833. *Principles of Geology: Being an Attempt to Explain the Former Changes in the Earth's Surface by Reference to Causes now in Operation*. London: John Murray.

Paterniani, E. 1969. Selection for reproductive isolation between two populations of maize, *Zea mays* L. *Evolution* 23: 534–47.

Popper, K. R. 1959. *The Logic of Scientific Discovery*. London: Hutchinson.

Popper, K. R. 1972. *Objective Knowledge: An Evolutionary Approach*. Oxford: Oxford University Press.

Popper, K. R. 1974. Darwinism as a metaphysical research programme. In *The Philosophy of Karl Popper* Vol. 1 (Editor P. A. Schilpp ), 133–43. LaSalle, IL: Open Court.

Reid, T. [1785] 2002 *Essays on the Intellectual Powers of Man*. (Editor D. Brookes ) University Park: Pennsylvania State University Press.

Ruse, M. 1975b. Darwin's debt to philosophy: an examination of the influence of the philosophical ideas of John F. W. Herschel and William Whewell on the development of Charles Darwin's theory of evolution. *Studies in History and Philosophy of Science* 6: 159–81.

Ruse, M. 1981. What kind of revolution occurred in geology? *PSA 1978*, 2: 240–73. East Lansing: Philosophy of Science Association.

Ruse, M. 2005b. Darwin and mechanism: Metaphor in science. *Studies in History and Philosophy of Biology and Biomedical Sciences* 36: 285–302.

Sober, E. 2011. *Did Darwin Write the "Origin" Backwards? Philosophical Essays on Darwin's Theory*. Buffalo, NY: Prometheus.

von Humboldt, A. 1814. *Personal Narrative of Travels of the Equinocial Regions of the New Continent during Years 1799–1804*. London.

Whewell, W. 1837. *The History of the Inductive Sciences (3 vols)*. London: Parker.

Whewell, W. 1840. *The Philosophy of the Inductive Sciences (2 vols)*. London: Parker.

## Chapter 6

Adams, M. 2021. Little Evolution, BIG Evolution: Rethinking the History of Darwinism, Population Genetics, and the "Synthesis". *Natural Selection: Revisiting its Explanatory Role in Evolutionary Biology*. (Editor R. G. Delisle), 195–230. Cham, Switzerland: Springer.

Ceci, S. J. and W. M. Williams. 2009. *The Mathematics of Sex: How Biology and Society Conspire to Limit Talented Women and Girls*. New York: Oxford University Press.

Cosmides, L. and J. Tooby. 1997. *Evolutionary Psychology: A Primer*. http/wwwpysch.ucsb.edu/research/cep/primer.html.

Dart, R. 1953. The predatory transition from ape to man. *International Anthropological and Linguistic Review* 1.4: 201–17.

Eldredge, N. and S. Gould 1972. Punctuated equilibria: an alternative to phyletic gradualism. *Models in Paleobiology*. (Editor T. J. M. Schopf ) 82–115. San Francisco: Freeman, Cooper.

Fry, D. P. 2014. Group Identity as an Obstacle and Catalyst of Peace.*Pathways to Peace: The Transformative Power of Children and Families*. (Editors J. F. Leckman, C. Panter-Brick, and R. Salah ) 79–92. Cambridge, MA: MIT Press.

Fry, D. P., C. A. Keith, and P. Söderberg. 2020. Social complexity, inequality and war before farming: congruence of comparative forager and archaeological data.*Social Inequality before Farming. Multidisciplinary approaches to the study of social organization in prehistoric and ethnographic hunter gatherer-fisher societies*. (Editor L. Moreau ) 303–20. Cambridge: McDonald Institute for Archaeological Research.

Grant, P. R. and R. B. Grant. 2007. *How and Why Species Multiply: The Radiation of Darwin's Finches*. Princeton, NJ: Princeton University Press.

Hamilton, W. D. 1964. The genetical evolution of social behaviour. *Journal of Theoretical Biology* 7: 1–32.

Hull, D. L. 1980. Individuality and selection. *Annual Review of Ecology and Systematics* 11: 311–32.

Lieberman, D. E. 2013. *The Story of the Human Body: Evolution, Health, and Disease*. New York: Vintage.

MacArthur, R. H. and E. O. Wilson. 1967. *The Theory of Island Biogeography*. Princeton: Princeton University Press.

Reznick, D. N., J. Losos, and J. Travis. 2018. From low to high gear: there has been a paradigm shift in our understanding of evolution. *Ecology Letters* 22: 233–44.

Ruse, M. 1979b. *Sociobiology: Sense or Nonsense?* Dordrecht, Holland: Reidel.

Ruse, M. 2022. *Why We Hate: Understanding the Roots of Human Conflict*. Oxford: Oxford University Press.

Segerstrale, U. 2000. *Defenders of the Truth: The Battle for Science in the Sociobiology Debate and Beyond*. New York: Oxford University Press.

Sepkoski Jr, J. J. 1979. A kinetic model of Phanerozoic taxonomic diversity. II Early Paleozoic families and multiple equilibria. *Paleobiology* 5: 222–52.

Sepkoski Jr, J. J. 1984. A kinetic model of Phanerozoic taxonomic diversity. III Post-Paleozoic families and mass extinctions. *Paleobiology* 10: 246–67.

Simpson, G. G. 1944. *Tempo and Mode in Evolution*. New York, NY: Columbia University Press.

West, S. A., A. S. Griffin, and A. Gardner. 2007. Social semantics: altruism, cooperation, mutualism, strong reciprocity and group selection. *Journal of Evolutionary Biology* 20: 415-32.

Wilson, E. O. 1975. *Sociobiology: The New Synthesis*. Cambridge, Mass.: Harvard University Press.

## Chapter 7

Allen, E. et al. 1976. Sociobiology: a new biological determinism. *BioScience* 26: 182–86.

Allen, C. E., P. Beldade, B. J. Zwann, and P. Brakefield. 2008. Differences in the selection response of serially repeated color pattern characters: standing variation, development, and evolution. *BMC Evolutionary Biology* 8: 94.

Arthur, W. 2021. *Understanding Evo-Devo*. Cambridge: Cambridge University Press.

Beldade, P., K. Koops, and P. M. Brakefield. 2002. Developmental constraints versus flexibility in morphological evolution. *Nature* 416: 844–47.

Conway Morris, S. 2003. *Life's Solution: Inevitable Humans in a Lonely Universe*. Cambridge: Cambridge University Press.

Coyne, J. A., N. H. Barton, and M. Turelli 1997. Perspective: a critique of Sewall Wright's shifting balance theory of evolution. *Evolution* 51: 643–71.

Darwin, C. 1859. *On the Origin of Species by Means of Natural Selection, or the Preservation of Favoured Races in the Struggle for Life*. London: John Murray.

Darwin, C. 1861. *Origin of Species* 3rd Ed.London: John Murray.

Darwin, C. 1987. *Charles Darwin's Notebooks, 1836–1844*. (Editors P. H. Barrett, P. J. Gautrey, S. Herbert, D. Kohn, and S. Smith ) Ithaca, NY: Cornell University Press.

Dawkins, R. 1986. *The Blind Watchmaker*. New York, NY: Norton.

Dupré, J. 2003. *Darwin's Legacy: What Evolution Means Today*. Oxford: Oxford University Press.

Dupré, J. 2010. The conditions for existence. *American Scientist* 98: 170.

Dupré, J. 2012a. *Processes of Life: Essays in the Philosophy of Biology*. Oxford: Oxford University Press.

Dupré, J. 2012b. Evolutionary theory's welcome crisis. *Project Syndicate* https://www.project-syndicate.org/commentary/evolutionary-theory-s-welcome-crisis-by-john-dupre?barrier=accesspaylog

Dupré, J. 2017. The metaphysics of evolution. *Interface Focus* https://doi.org/10.1098/rsfs.2016.0148

Fodor, J. 2007. Why pigs don't have wings. The case against natural selection. *London Review of Books* 29: 18 October.

Fodor, J. and M. Piattelli-Palmarini. 2010. *What Darwin Got Wrong*. New York: Farrar, Straus, and Giroux.

Futuyma, D. J. 2017. Evolutionary biology today and the call for an extended synthesis. *Interface Focus* https://doi.org/10.1098/rsfs.2016.0145

Gilbert, S. F., J. M. Opitz, and R. A. Raff. 1996. Resynthesizing evolutionary and developmental biology. *Developmental Biology* 173: 357–72.

Gish, D. 1973. *Evolution: The Fossils Say No!* San Diego: Creation-Life.

Gould, S. J. and R. C. Lewontin. 1979. The spandrels of San Marco and the Panglossian paradigm: a critique of the adaptationist programme. *Proceedings of the Royal Society of London, Series B: Biological Sciences* 205: 581–98.

Haldane, J. B. S. 1927. *Possible Worlds and Other Essays*. London: Chatto and Windus.

Haldane, J. B. S. 1932. *The Causes of Evolution*. New York: Cornell University Press.

Kimura, M. 1968. Evolutionary rate at the molecular level. *Nature* 217: 624–6.

Laland, K. , J. Odling-Smee, and J.Endler, 2017. Niche construction, sources of selection and trait coevolution. *Interface Focus* https://royalsocietypublishing.org/doi/10.1098/rsfs.2016.0147

Laland, K. N., T. Uller, M. W. Feldman, et al. 2015. The extended evolutionary synthesis: its structure, assumptions and predictions. *Proceedings of the Royal Society B*. https://kevintshoemaker.github.io/EECB-703/Laland%20et%20al.%20-%202015%20-%20The%20extended%20evolutionary%20synthesis%20its%20structure.pdf

Lewontin, R. C. 2002. *The Triple Helix: Gene, Organism, and Environment*. Cambridge, MA: Harvard University Press.

McShea, D. W. and R. Brandon. 2010. *Biology's First Law: The Tendency for Diversity and Complexity to Increase in Evolutionary Systems*. Chicago: University of Chicago Press.

Nagel, T. 2012. *Mind and Cosmos: Why the Materialist Neo-Darwinian Conception of Nature Is Almost Certainly False*. New York: Oxford University Press.

Owen, R. 1849. *On the Nature of Limbs*. London: Voorst.

Reznick, D., M. J. Butler, and H. Rodd. 2001. Life-history evolution in guppies. VII. The comparative ecology of high- and low-predation environments. *American Naturalist* 157: 12–26.

Richards, R. J. 2003. *The Romantic Conception of Life: Science and Philosophy in the Age of Goethe*. Chicago: University of Chicago Press.

Richards, R. J. 2013. *Was Hitler a Darwinian? Disputed Questions in the History of Evolutionary Theory*. Chicago: University of Chicago Press.

Ruse, M. 1996. *Monad to Man: The Concept of Progress in Evolutionary Biology*. Cambridge, Mass.: Harvard University Press.

Vermeij, G. J. 1987. *Evolution and Escalation*. Princeton, NJ: Princeton University Press.

Waddington, C. H. 1960. *The Ethical Animal*. London: Allen and Unwin.

Whitehead, A. N. 1926. *Science and the Modern World*. Cambridge: Cambridge University Press.

Williams, G. C. 1966. *Adaptation and Natural Selection*. Princeton, NJ: Princeton University Press.

## Chapter 8

Bergson, H. 1911. *Creative Evolution*. New York: Holt.

Bradley, B. 2020. *Darwin's Psychology*. Oxford: Oxford University Press.

Brewer, M. B. 1999. The Psychology of Prejudice: Ingroup Love or Outgroup Hate? *Journal of Social Issues* 55: 429–44.

Carnegie, A. 1889. The Gospel of Wealth. *North American Review* 148: 653–65.

Crook, P. 1994. *Darwinism, War and History: The Debate over the Biology of War from the 'Origin of Species' to the First World War*. Cambridge: Cambridge University Press.

Dawkins, R. 2006. *The God Delusion*. New York: Houghton, Mifflin, Harcourt.

Darwin, C. 1871. *The Descent of Man, and Selection in Relation to Sex*. London: John Murray.

Fitzgerald, F. S. 1945. *The Crack-Up*. New York: New Directions.

Frost, R. 1931. Education by poetry: a meditative monologue. *Amherst Graduates' Quarterly* XX: 75–85.

Gibson, A. 2013. Edward O. Wilson and the organicist tradition. *Journal of the History of Biology* 46: 599–630.

Gould, S. J. 1977. *Ontogeny and Phylogeny*. Cambridge, MA: Belknap Press.

Gould, S. J. 1988. On replacing the idea of progress with an operational notion of directionality. In *Evolutionary Progress*. (Editor M. H. Nitecki) 319–38. Chicago: University of Chicago Press.

Gould, S. J. and R. C. Lewontin. 1979. The spandrels of San Marco and the Panglossian paradigm: a critique of the adaptationist programme. *Proceedings of the Royal Society of London, Series B: Biological Sciences* 205: 581–98.

Grant, B. R. and P. R. Grant. 1993. Evolution of Darwin's finches caused by a rare climatic event. *Proceedings of the Royal Society: Biological Sciences* 251: 111–17.

Grant, P. R. and B. R. Grant. 1995. Predicting microevolutionary responses to directional selection on heritable variation. *Evolution* 49: 241–51.

Grant, B. R. and P. R. Grant. 2003. What Darwin's finches can teach us about the evolutionary origin and regulation of biodiversity. *BioScience* 53: 965–75.

Hegel, G. W. F. [1821] 1991. *Elements of the Philosophy of Right*. (Editor A. Wood ) Cambridge: University of Cambridge Press.

Hitler, A. [1925] 2009. *Mein Kampf*. (Translator M. Ford) http://www.hitler-library.org/

Hume, D. [1739–1740] 1978. *A Treatise of Human Nature*. Oxford: Oxford University Press.

Huxley, T. H. 1893. *Evolution and Ethics with a New Introduction*. (Editor M. Ruse) Princeton: Princeton University Press.

Jablonski, N. G. 2004. The evolution of human skin and skin color. *Annual Review of Anthropology* 33: 585–623.

Johnson, G. R. 1986. Kin selection, socialization, and patriotism: an integrating theory. *Politics and the Life Sciences* 4: 127–40.

Matthen, M. and A. Ariew. 2009. Selection and causation. *Philosophy of Science* 76: 201–24.

O'Connell, J. and M. Ruse 2021. *Social Darwinism (Cambridge Elements on the Philosophy of Biology)*. Cambridge: Cambridge University Press.

Pence, C. H. 2021. *The Causal Structure of Natural Selection*. Cambridge: Cambridge University Press.

Popper, K. R. 1974. Darwinism as a metaphysical research programme. In *The Philosophy of Karl Popper* Vol. 1. (Editor P. A. Schilpp ), 133–43. LaSalle, IL: Open Court.

Reich, D. 2018. *Who We Are and How We Got Here: Ancient DNA and the New Science of the Human Race*. New York: Pantheon.

Reynolds, A. S. 2022. *Understanding Metaphors in the Life Sciences*. Cambridge: Cambridge University Press.

Richards, R. J. 2013. *Was Hitler a Darwinian? Disputed Questions in the History of Evolutionary Theory*. Chicago: University of Chicago Press.

Spencer, H. 1870. On ancestor worship and other peculiar beliefs. *Fortnightly Review* 13: 535–50.

Von Bernhardi, F. 1912. *Germany and the Next War*. London: Edward Arnold.

Wallace, A. R. 1858. On the tendency of varieties to depart indefinitely from the original type. *Journal of the Proceedings of the Linnean Society, Zoology* 3: 53–62.

Weikart, R. 2013. The role of Darwinism in Nazi racial thought. *German Studies Review* 36: 537–56.

# Figure Credits

| | |
|---|---|
| Figure 1.2 | Drawn by Martin Young. |
| Figure 1.3 | Reproduced from Darwin's Notebook, B, July 1837. |
| Figure 1.4 | Reproduced from *On the Origin of Species* (1859) by Charles Darwin. |
| Figure 1.5 | Drawn by Martin Young. |
| Figure 1.6 | Reproduced from *Palaeontology* (1860) by Richard Owen. |
| Figure 1.7 | Reproduced from *On the Nature of Limbs* (1849) by Richard Owen. |
| Figure 2.1 | Reproduced from the woodcut *Ladder of Ascent and Descent of the Mind* (1305) by Raymond Lull. |
| Figure 4.1 | Drawn by Martin Young. |
| Figure 4.2 | Drawn by Martin Young. |
| Figure 4.3 | a, Taken from https://en.wikipedia.org/wiki/File:Biston.betularia.7200.jpg (made available under a Creative Commons Attribution-Share Alike 3.0 Unported licence). b, Taken from https://en.wikipedia.org/wiki/File:Biston.betularia.f.carbonaria.7209.jpg (made available under a Creative Commons Attribution-Share Alike 2.5 Generic licence). |
| Figure 4.4 | Drawn by Martin Young. |
| Figure 5.1 a, b | Reproduced from *The Structure and Distribution of Coral Reefs* (1842) by Charles Darwin. |
| Figure 5.2 | Reproduced from *Observations on the parallel roads of Glen Roy* (1839) by Charles Darwin. |

Figure 5.3      Mayr, E. 1942. *Systematics and the Origin of Species.* New York, NY: Columbia University Press p. 183.

Figure 6.1      Adapted by Martin Young from *The Theory of Island Biogeography* (1967) by Robert H. MacArthur and Edward O. Wilson. Reprinted with permission from Princeton University Press.

Figure 6.2      Reproduced from Sepkoski Jr, J. J. (1984). A kinetic model of Phanerozoic taxonomic diversity. III Post-Paleozoic families and mass extinctions. *Paleobiology* 10: 246–67. Reprinted with permission from Cambridge University Press.

Figure 6.3      Courtesy of Jonathan Haas.

Figure 7.1      Drawn by Martin Young.

Figure 7.2      From Gould, S. J., and R. C. Lewontin. 1979. The spandrels of San Marco and the Panglossian paradigm: a critique of the adaptationist programme. *Proceedings of the Royal Society of London, Series B: Biological Sciences* 205: 581–98.

Figure 7.3 a, b    Reproduced from *Darwinism and its Discontents* (2006) by Michael Ruse. Reprinted with permission from Cambridge University Press.

Figure 7.4      Reproduced from *Understanding Evo-Devo* ©Wallace Arthur/Cambridge University Press 2021 with permission.

Figure 8.1      Reproduced from *The Evolution of Man* (1896) by Ernst Haeckel.

# Index